DATE DUE			
Oct 7 78			
Nov 19 78			
Feb 13 79			

RESEARCH IN ZOOS AND AQUARIUMS

A symposium held at the
Forty-ninth Conference of the
AMERICAN ASSOCIATION
OF ZOOLOGICAL
PARKS AND AQUARIUMS
Houston, Texas
October 6–11, 1973

INSTITUTE OF LABORATORY
ANIMAL RESOURCES
Assembly of Life Sciences
National Research Council

NATIONAL ACADEMY OF SCIENCES
WASHINGTON, D.C.   1975

This publication was supported with primary support from Contract PH43-64-44 with the Drug Research and Development, Division of Cancer Treatment, National Cancer Institute, and the Animal Resources Branch, National Institutes of Health; and partial support from Contracts N01-CP-33338 and N01-CP-45617 with the Division of Cancer Cause and Prevention, National Cancer Institute, National Institutes of Health, U.S. Public Health; Contract AT(11-1)-3369 with the Atomic Energy Commission; Contract N00014-67-A-0244-0016 with the Office of Naval Research, U.S. Army Medical Research and Development Command, and U.S. Air Force; Contract 12-16-140-155-91 with the Animal and Plant Health Inspection Service, U.S. Department of Agriculture; Contract NSF-C310, Task Order 173, with the National Science Foundation; Grant RC-1Q from the American Cancer Society, Inc.; and contributions from pharmaceutical companies and other industry.

Library of Congress Cataloging in Publication Data
Main entry under title:

Research in zoos and aquariums.

Includes bibliographies.
1. Zoology, Experimental—Congresses. 2. Zoo animals—Congresses. 3. Zoological gardens—Congresses.
I. American Association of Zoological Parks and Aquariums.
QL1.R27   596'.007'24   75-4962
ISBN 0-309-02319-X

Available from
Printing and Publishing Office, National Academy of Sciences
2101 Constitution Avenue, N.W., Washington, D.C. 20418

Printed in the United States of America

596.00724
R31
106197
Sept. 1978

Preface

This symposium was developed by the Institute of Laboratory Animal Resources in cooperation with the American Association of Zoological Parks and Aquariums and held on the occasion of the forty-ninth annual conference of that Association. As such, it represents the first such major symposium held in conjunction with the annual conference, and was intended to focus on research in and by zoos. It was intended, further, to promote better understanding and closer cooperation between those whose interests and experience lie with zoos and their counterparts in the biomedical and academic research community.

The individual papers, presented in a series of four topically oriented sessions, generally support the view that zoos and aquariums represent a large, but primarily unutilized, potential for scientific investigations. The first session focused on general aspects of administration, funding, and research opportunities; the second dealt with behavioral research; the third stressed reproductive biology; the fourth cited special applications. A summary and perspective was provided by Dr. Cluff E. Hopla, chairman of ILAR, who shared with Dr. Lester Fisher, president of AAZPA, overall responsibility for planning the symposium.

Contents

I

GENERAL ASPECTS

KURT BENIRSCHKE, M.D.
Department of Reproductive Medicine
University of California, San Diego

Biomedical Research

The presence within zoos and aquariums of many exotic species has tempted numerous investigators. Here, they conjecture, lie opportunities totally unexploited to explain relationships to man, to investigate diseases, to seek models that might be explored to better understand human conditions or biological phenomena. As a result, many a scientist and practicing physician from the outside has found his way inside to partake of this wealth of material. He has collaborated with zoo pathologist, veterinarian, or curator to obtain samples to further his understanding of a given phenomenon. Many "outsiders," however, have found this a frustrating experience for—much to their chagrin—animals in collections would not be manipulated at will, refusing to be immobilized with ease or at a specified time; the already overburdened zoo personnel could not help; the investigator was summoned to the death of his favorite creature at an inconvenient time, and so forth. Often the investigator and zoo personnel were unable to communicate their respective problems adequately and much frustration ensued.

Thus, while outside investigators have conducted major research projects within zoos or through their help, it has been a sporadic effort and almost always on questions that have little direct relevance to the collection itself. Most zoo personnel are acquainted with this aspect of the overall problem of research in zoos. Requests in zoo pathologists' morgues are long for tissues, blood, serum, carcass, hair, feathers, and so

Support by a grant from The Rockefeller Foundation (RF70029) is gratefully acknowledged.

3

forth. Zoo personnel have a difficult time sorting these requests and, even with good will, cannot comply with them all. Yet such specimens are vital to further knowledge, and systematic questions have been answered in the past by their supply. In addition to requests, other forms of outside research include short visits by investigators when animals are immobilized, transported or treated and the longer term visits of scientists-in-residence, usually excused from university duties during sabbatical. The most gratifying experiences of this kind have been with this latter type of association: First, the arrival of the scientist-in-residence is preceded by a period of prolonged negotiation that settles the questions of space, equipment, availability of animals, and so forth; second, the investigator is in residence for a time that is sufficiently long to get accustomed to the special circumstances of the zoo environment. Often, the scientist participates in the chores of handling, transport, special feeding, and caring for the animals to be studied and an improved rapport develops between him and the zoo staff. Finally, and most important for the achievement of specific goals, scientists on sabbatical not only bring much new knowledge to the zoo staff through their intimate interactions but the investigations also have immediate relevance to the collection.

Another important aspect is the in-house research performed as an ongoing project by park staff. To stimulate further development of this aspect has been one reason for holding this symposium. In a paper entitled "The Value of Zoos for Science and Conservation," Caroline Jarvis (1967) stated that, "Zoos have an unfortunate record in conservation and research. . . . However, the concept of the Zoo as a place of research and education, rather than a public spectacle and entertainment, is only just beginning to be established even in the scientific world, let alone in the minds of the general public." She goes on to give many examples of where meritorious activities might be conducted.

Another effort to review the research potential in zoological gardens has been made by Schroeder (1970). His sad conclusions can be summarized as follows, "The research effort to date has been puny, and for the most part studies which might give greatest reward have not been pursued."

These and other reviews are exemplary of the discouragement that is generally expressed when research at zoos is discussed and, I am afraid, many of my scientific colleagues have turned away from even trying to exploit the riches that parks offer to their investigative pursuits.

Collecting data for in-house research in zoos is difficult for a number of reasons. Many curators, veterinarians, and pathologists employed by the zoos pursue research work that leads to publication; however, because it is so widely scattered over the literature, it cannot be meaningfully summarized. Nor is it a generally sustained, well-identified activity; most employees of parks have duties that allow them only part time for research. In only a few zoos is there a well-identified separate research division where the staff has few or no responsibilities for the maintenance of the collection and are employed for the sole purpose of investigation. Best known in this respect, of course, is the Nuffield Institute, associated with the London Zoological Society, with its enviable record of "basic research" in many aspects of reproduction, primatology, and other biomedical problems. In this country few institutions have a longer-standing research effort—New York, Philadelphia, Washington, San Diego; and, perhaps, others with less broadly conceived charges. However, considering the large number of zoological gardens and aquariums in the United States that contain a large number of animals, the overall research is indeed slight. Moreover, the fact that these animals are involuntarily captive and subjected to stresses, infections, and other unwanted exigencies is warrant enough for greater efforts to improve their health, mental well-being, reproductive success, caging, and nutrition.

Since Dr. Conway will describe the research activities in New York, I will restrict myself to a brief view of the other institutions mentioned. The Penrose Laboratories at Philadelphia have espoused basic research for many years with admirable success. Infections and nutritional diseases of a wide variety of animals have been the focuses of attention and, over the years, this group has pioneered in understanding the effects of crowding and disease. Work conducted there on woodchucks, mice, chickens, and other animals led to a much better understanding of adrenal physiology, sudden death, arteriosclerosis, and so forth. Many of these findings have been summarized in two papers at symposia that have dealt specifically with research within zoos (Snyder and Ratcliffe, 1969; Snyder, 1967). One problem with Philadelphia's research effort has been that there was never what might be called a "critical mass" of investigators; this emphasizes the need for strong ties to universities. The National Zoological Park at Washington, D.C., on the other hand, has strong research affiliations with the Armed Forces Institute of Pathology and National Institutes of Health, and some really important work has taken place there.

A systematic effort is being made at the National Zoological Park to arrive at a means via the endocrine system for early detection of pregnancy in apes. The excretion of chorionic gonadotropin and estrogens in the pregnant gorilla, particularly the finding of estriol in pregnancy, have important comparative value (Tullner and Gray, 1968; Hopper, Tullner, and Gray, 1968). Lead poisoning in primates (Zook *et al.*, 1973) has important biomedical connotations. As in many other zoos, geneticists have reaped a rich harvest from the wide variety of species available. Evolutionary relationships can be studied by analysis of amino acids from primates and other orders. At Washington the study of unusual hemoglobins in apes is giving valuable direction to the study of heterogeneous types of human thalassemias (Boyer *et al.,* 1971). Other examples can be cited where an in-house research group at the Washington zoo, in collaboration with investigators from other institutions, has made valuable contributions to biomedical progress.

At San Diego, research is one of the charges in the charter of the Zoological Society, and a zoo hospital was built in 1927. In addition to housing the pathology staff and veterinarians, it is well equipped for research. Over the years this hospital has been occupied by many investigators, and the Scripps Foundation has sponsored fellowships since 1937. Research on toxicology, lead poisoning in birds, reproduction, parasitology, coagulation problems in primates, bilirubin transport, blood grouping, comparative oxygen dissociation of hemoglobins, and many other problems of a biomedical nature has been conducted here by visitors, often in close collaboration with staff. However, despite repeated attempts a continuously successful in-house research operation has not been achieved. New efforts are currently under way towards its creation. Jarvis (1967) mentions the meritorious work in virus pathology, research that has regrettably ceased. While active, it yielded much new information on herpes viruses, on the nature of oncogenic viruses, and had important by-products for the zoo. It gave insight into reproductive biologic parameters of new world monkeys and talapoins, defined nutritional requirements of these species, and brought into focus some of the possibilities of research within zoos. At the same time, it has pointed to existing problems referred to at the end of this discussion.

Much research conducted internally in zoos has had wide biomedical ramifications, but perhaps none more than that conducted by pathologists. Comparative pathology has grown to be a respected field and research in this area has resulted in the discovery of many animal models for studying human diseases. An extensive file including numerous publications on this subject exists at San Diego. Not all of the diseases on file are infectious in nature—many are of genetic interest and of direct pertinence to man. The Institute of Laboratory Animal Resources (ILAR) also has a registry that collects data of this type for dissemination to the scientific community. Knowledge of animal models has been helpful in the study of a variety of human conditions.

Biomedical research in zoos has benefited man, but many examples can also be cited where research in man becomes directly applicable to animal collections. At a previous occasion, the celebration of Antwerp Zoo's one hundred twenty-fifth anniversary, I have reviewed some aspects of this two-way street between the investigator and the zoo. I want to give one further example from my own experience in order to emphasize the need for research in zoos.

In the last twenty years knowledge of the mammalian genome has advanced enormously through cytogenetics, primarily through the recognition that chromosome errors are a frequent cause of disease in man, the discovery of simple techniques, and extensive work in numerous universities and hospitals. Chromosomal errors in man are the cause of most abortions, of many types of mental deficiency, some types of infertility, some neoplasias, and so forth. It is now known that 0.5 percent of newborns have an abnormal chromosome number (Jacobs, 1971). Whether the same is true in other mammals is not fully understood as yet; however, many similar conditions have been discovered in domestic species and some zoo animals. For example, the sterility of male calico cats can be explained by chromosome errors, and there is reason to believe that many other examples will come to light upon systematic inquiry. In the normal animal population, such systematic study is being conducted by a few investigators and cooperation with a few parks has been of enormous help in advancing knowledge of the genetic relationship of various taxa. Recently, through the development of new "banding" techniques, it has become possible to take an entirely new look at these aspects of chromosomal evolution. Findings have come to light that make it mandatory for zoos to not only be knowledgeable but to participate more actively in this area: For example, at least three subspecies of squirrel monkeys exist that are chromosomally very distinct (Ma *et al.,* 1974) and whose interbreeding in zoos should be prohibited. In spider monkeys, likewise, taxonomic delineation has been difficult. Through collaboration with Dr. R. A. Cooper, who collected samples from distinct and isolated groups of *Ateles* in Colombia, similar differences could be identified. In this species, chromosome number 1 takes two distinctly different forms (Figure 1): *Ateles paniscus belzebuth,* the Amazonian variety, has a submetacentric element; *A. p. robustus* from Cordoba possesses a more metacentric first chromosome. The change has come about through a pericentric inversion at points indicated by arrows. Other types of chromosomal polymorphism in species are now known. When such species are propagated in zoos these aspects must be understood lest hybrids with possible infertility be produced. For biomedicine the recognition in mammals of such inversions will be invaluable, for they are recognized in man. Here, reproductive failure is at times associated with such inversion in its heterozygous state, but difficulties present themselves. These species may possibly lend themselves well to further exploration to the benefit of both parties. Such has been the case in muntjacs. Of the two species in zoos, Reeve's muntjac

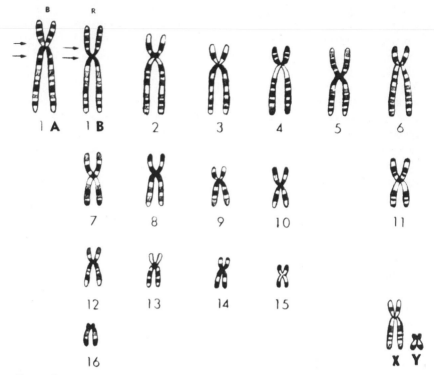

Figure 1

Schematic diagram of the Giemsa-banding pattern of chromosomes of two forms of spider monkey. Chromosome 1A is the type found in *Ateles paniscus belzebuth* (collected in Bolivar, Venezuela), 1B is the type representative of *A. p. robustus* (collected in Cordoba, Argentina by Dr. R. A. Cooper). The arrows indicate break points of a segment that has undergone pericentric inversion. No difference could be identified in the remaining chromosomes.

was found to have 46 chromosomes and the Indian muntjac only 7 in the male and 6 in the female (Wurster and Benirschke, 1970). Species designation is easy by this means, and one can safely say that hybridization is unlikely or at least would yield infertile specimens. In this case biology has also benefited; the recognition in a mammal of such a low chromosome number has enabled scientists to make inroads into mammal chromosome structure, segregation mechanics, translocation events, and the like—aspects formerly reserved to insect cytogenetics.

It is idle to catalogue further achievements of relevant biomedical research conducted in zoos. The reason for this symposium is that we are unhappy about the magnitude, if not the excellence, of research within parks. If the zoo is a storehouse for scientific investigations, why then is this storehouse not more uniformly exploited? In my opinion, it is primarily because it is difficult to "sell" the idea of basic research.

As this brief review has shown, it is difficult to define "basic" research, since usually its findings are soon "applied." To be sure, the area of application of basic research is difficult to forecast and often it is totally unexpected. Conversely, when research is conducted solely with the purpose of yielding immediate "applied" results, it has frequently been less rewarding and, to the investigator, less stimulating. Rather than discussing the semantics of "applied versus basic," we should address ourselves to the excellence and magnitude of research endeavors. It is unlikely that most of the important problems of zoos and aquariums will be tackled or settled by the sporadic visitations of outside investigators; their solution must come from in-house research. This symposium testifies to the acceptance of this dictum by at least some responsible persons. How is it best implemented?

I have had the opportunity to prepare for the Board of Trustees of the San Diego Zoological Society a white paper on research which, I hope, will soon yield results. This review of the topic contains the following principal suggestions:

1. One or more principal research investigators should be hired by the Society.
2. Strong direct affiliations to the local universities should be established.
3. Inside and outside funding should be secured.
4. Collaboration with the scientific community should be expanded by the solicitation of graduate students and other outside investigators to come for short-term work and by the enhancement of our local tradition of professors coming for sabbaticals.

I believe that all major zoos will have to acquire some scientists whose principal responsibility is research, and that each zoo will have its own preferences in the type of work to be accomplished. An overriding need exists in all zoos for a better understanding of reproduction; likewise, zoos offer great opportunities for research in all kinds of reproductive physiology, and associated endocrine, psychological, and genetic studies. Only London can claim to have established an institute that might serve as a model for creative work in areas that can only be pursued alongside the collection and with the collaboration of curators, veterinarians, and other zoo staff. In San Diego, behavioral and microbiological investigation will be the focus of future studies, studies that are undeniably essential for reproductive success in zoos. In addition, we hope to establish a cell and sperm bank. The commitment for these activities will have to come from the parks or the zoological societies of the United States, lest the long-term achievements be jeopardized by the whims of granting agencies.

Much can be accomplished by strong ties to the universities that make it possible to attract the most qualified investigators. Here, the efforts are rudimentary but they must be pushed to arrive at solutions of a practical nature. One possible association would be to establish endowed chairs in universities for investigators who carry on their work in the zoological parks. A number of benefits can accrue to the zoo and the university through such direct affiliation. Not only will the quality of researcher probably be better but, through his interactions with students and faculty, a continuous input of young people to the site of research can also be anticipated. Further, scientific publications will enhance the quality of work. Many other types of association can be envisaged; however, it is important first that the imaginary fence erected between the two parties be taken down.

Funding for research must be derived in part from zoo sources, at least in those institutions whose charters stipulate research. Once firmly established, a wide variety of research support can be tapped, providing of course that the work has sufficient excellence and scope. Numerous conservation organizations, private foundations, the government, and also the public are susceptible to proper appeals. It is important to reiterate that these outside agents should not be relied on to form the nucleus of zoo research support.

Finally, a warm hand must be extended from the zoos to those in the scientific community who have an intense desire to partake in their animal wealth. And the outside investigator must understand the restrictions that are peculiar to zoo animals with respect to capture, immobilization, blood sampling, and the like. Too often trivial problems of misunderstanding have thwarted efforts at collaboration. The presence of investigators within the zoos should do much to enhance cooperation in this respect, to the benefit of both.

REFERENCES

Benirschke, K. 1969. Zoos and the pathologist—A two-way street, or, cytogenetics on zoo animals. Acta Zool. Pathol. Antverpiensia 48:29–41.

Boyer, S. H., A. N. Noyes, G. R. Vrablik, L. J. Donaldson, E. W. Schaefer, C. W. Gray, and T. F. Thurmon. 1971. Silent hemoglobin alpha genes in apes: Potential source of thalassemia. Science 171:182–185.

Brown, T. M., H. W. Clark, J. S. Bailey, and C. W. Gray. 1970. A mechanistic approach to treatment of rheumatoid type arthritis naturally occurring in a gorilla. Trans. Clin. Chimatol. Assoc. 82:227–247.

Hopper, B. R., W. W. Tullner, and C. W. Gray. 1968. Urinary estrogen excretion during pregnancy in a gorilla. Proc. Soc. Exp. Biol. Med. 129:213–214.

Jacobs, P. A. 1971. Chromosome abnormalities and fertility in man. Pages 346–358 in R. A. Beatty and S. Gluecksohn-Waelsch, eds. The genetics of the spermatozoon. Proceedings of an International Symposium, Bogtrykkeriet Forum, Copenhagen.

Jarvis, C. 1967. The value of zoos for science and conservation. Oryx 9:127–136.

Ma, N. S. F., T. C. Jones, R. W. Thorington, and R. W. Cooper. 1974. Chromosome banding patterns in squirrel monkeys (*Saimiri sciureus*). J. Med. Primatol. (in press)

Schroeder, C. R. 1970. Research potential in zoological gardens. Zool. Garten 39: 240–247.

Snyder, R. L. 1967. Fertility and reproductive performance of grouped male mice. Pages 458–472 in K. Benirschke, ed. Comparative aspects of reproductive failure. Springer-Verlag, New York.

Snyder, R. L., and H. L. Ratcliffe. 1969. *Marmota monax*: A model for studies of cardiovascular, cerebrovascular and neoplastic disease. Acta Zool. Pathol. Antverpiensia 48:265–273.

Tullner, W. W., and C. W. Gray. 1968. Chorionic gonadotropin excretion during pregnancy in a gorilla. Proc. Soc. Exp. Biol. Med. 128:954–956.

Wurster, D. H., and K. Benirschke. 1970. Indiana muntjac, *Muntiacus muntjak*: A deer with a low chromosome number. Science 168:1364–1366.

Zook, B. C., R. M. Sauer, M. Bush, and C. W. Gray. 1973. Lead poisoning in zoo-dwelling primates. Am. J. Phys. Anthropol. 38:415–424.

JOHN F. EISENBERG

National Zoological Park
Smithsonian Institution, Washington, D.C.

Design and Administration of Zoological Research Programs

Although there is no sharp dividing line between what might be termed biomedical research and zoological research, it is a fact that these two research areas tend to develop autonomously within zoological institutions. At the National Zoological Park, research efforts are divided among three offices, including the Office of Pathology, the Office of Animal Health, and the Office of Zoological Research. Dr. Benirschke of San Diego has reviewed the biomedical aspects of research in zoological gardens. I will therefore confine my own discourse to the remaining research areas generally lumped under the loose title of "zoological studies" or a negative category—"*non*biomedical research."

Studies at major zoological gardens have historically included research on the behavior of animals, reproduction, and diets. Traditionally, much of the research in zoos was handled by the curatorial staff. Even when pressed with administrative duties, the dedication of administrative and curatorial staff members led to the production of many important contributions in the form of research papers (Schneider, 1930; Antonius, 1938) or books (Mann, 1930, Hediger, 1950, 1951; Crandall, 1964; Walker, 1964) to mention a few. There is also a long history of special institutions having established more defined areas of research responsibility, especially within those zoos founded by zoological societies. For example, the Penrose Laboratories were established in Philadelphia and charged at an early date with the responsibility of developing improved diets for specimens in the collection (Ratcliffe, 1936–1966). The zoo in Amsterdam is operated by the Royal Zoological Society, founded in 1838 with research laboratories established in 1928. Similar research

12

laboratories were established within the administrative framework of the Royal Zoological Society of Antwerp, the San Diego Zoological Society, the Chicago Zoological Society, the New York Zoological Society, and, of course, the Zoological Society of London. In more recent years, the Zoological Society of London expanded its research efforts with the establishment of the Wellcome Institution for research on vertebrate reproduction and the Nuffield Institute charged with biomedical research.

There are many unique factors, which have worked to produce a diversity of approaches to research problems for each major zoological garden in the United States and Europe. All of the research institutes associated with zoos have developed, however, by creating a core research staff, by providing an opportunity for guest investigators to rotate in attendance at the institute, and by rotation of postdoctoral or predoctoral university students. Almost all of the research institutes established in zoological gardens have some liaison with a local university. This relationship with universities and students cannot be stressed enough. The training of students fulfills a cultural mandate for any zoological institution worthy of the title. The coordination of such educational efforts demands some form of research staff; the educational value of such efforts has been duly documented by this Association.

In conjunction with activities of the institution itself, there is a tradition for sponsoring field research by the major zoological parks. Such field research in earlier years often consisted of extended trips by the curatorial staff to study specimens and procure them for a collection. Then, funding and operation of field stations developed, one of the most notable being the former research laboratories at Trinidad maintained for some years by the New York Zoological Society (Beebe, 1952). More recently, the Zoological Society of Frankfurt funded the Michael Grzimek Memorial Laboratories in what is now Tanzania. Such field efforts on a long-term basis encouraged feedback between field investigators and those curators charged with the maintenance of a captive collection.

The National Zoological Park is a bureau of the Smithsonian Institution and, as such, is in a unique position to involve itself in interbureau cooperation. The Smithsonian has its own system of field stations and laboratories, which are roughly comparable to an amalgam of the New York Zoological Society and the American Museum of Natural History. Both such complex systems are again comparable to our "parent" society in England, the Zoological Society of London.

In order to make my presentation relevant to zoological parks that are relatively autonomous units, and not part of such a complex system as is the case with New York, Washington, and London, I will attempt to outline the common problem areas that any zoological park faces and hopefully highlight solutions to some of the problems, i.e., the role of research in the zoological park, the selection of the staff, and funding.

SELECTION OF RESEARCH PRIORITIES

Research Goals

The major task in establishing a research unit involves first a determination of the roles for research personnel in the zoo and second the selection of the staff. Formulating areas of research responsibility necessitates priorities that may in part be unique to the founding institution, but the question of "pure" versus "applied" research will inevitably arise. Obviously the dichotomy between pure and applied research is in part false and the categories are not mutually exclusive. Above all, the zoological studies group must be integrated into the zoo itself and must be assigned an overall mission that is consistent with the philosophy of the institution. A research group must have the freedom to conduct so-called "pure" research problems because this encourages creativity and innovation. At the same time, the unit must have the flexibility to apply itself to practical problems within the park and the research effort should always involve aspects of public education that—at the least—should include participation by student trainees in the efforts of the unit.

So-called applied and non-applied projects may be combined in one study. A recent example included an analysis of social behavior in captive spider monkeys at the National Zoological Park, where the differential susceptibility to lead poisoning noted among primates in our collection was tied to species-specific behavioral traits (Zook, Eisenberg, and Mc-Lanahan, In press). At the same time analyses of auditory communication in the spider monkey and woolly monkey proceeded on schedule. So-called nonapplied and applied research projects are not incompatible, but require a sensitive understanding of management needs on the part of the research staff, as well as an appreciation of problems amenable to research on the part of the curatorial and maintenance staff.

At the outset, therefore, the development of the research program within a zoo should be divided in such a way that there are several missions including both applied and nonapplied research areas and that some means be maintained for disseminating information to the staff as a whole. Utmost care must be given to the selection of the core resident staff in order to obtain people temperamentally sympathetic to the overall aims of the zoological park.

Captive Propagation Programs

Zoos are becoming increasingly committed to conservation activities, and such conservation efforts may involve the captive propagation of endangered species. Sustained reproduction over many generations very often involves the application of research talents to the solution of management problems, as well as being an area in which researchers can make their most clear-cut contributions to a collection. For example, the National Zoological Park has committed itself to the maintenance of a golden marmoset (*Leontopithecus rosalia*) colony, and some of the research staff effort currently involves the analysis of bonding, the genesis of bonds, and the analysis of maturation patterns in the captive colony (Bridgwater, 1972). In a similar manner the captive propagation of the squirrel monkey (*Saimiri sciureus*) was pioneered by Cooper (1968) at San Diego. In Washington, we have also concerned ourselves with the analysis of the reproductive behavior shown by several ungulate species, including the Indian rhinoceros (*Rhinoceros unicornis*) and the sable antelope (*Hippotragus niger*) (Buechner, Stroman, and Xanten, 1975).

Research staffs are important to captive propagation programs and, indeed—with the Endangered Species Act in effect—many zoos will be unable to obtain import permits for such species unless they can demonstrate bona fide research efforts to promote sustained captive propagation. Thus, research staffs become pivotal if a zoo is going to develop a consistent in-house conservation program and become important in efforts within other countries undertaken in the cause of conservation. With respect to these latter efforts, the New York Zoological Society has been active for many years, and the creation of their Office of Field Studies is a logical extension of this tradition.

Field Research Efforts

Foreign countries may at times call upon zoological gardens for technical assistance. The National Zoological Park was involved for 4 years in the administration and supervision of grants for the survey of Ceylon's (Sri Lanka's) national parks (Buechner, 1968). These efforts were conducted jointly with the Office of Ecology, but the overall aims were recommendations for preservation of the wild elephant populations in Ceylon (McKay, 1973). This activity is extramural, but feedback into zoological parks is important. As a result of this study on elephants in Ceylon, material was published for the first time concerning the analysis of the reproductive cycle in the Asiatic elephant (Eisenberg, McKay, and Jainudeen, 1971).

SELECTION OF THE STAFF

The creation of a zoological research unit within a zoo requires that core personnel be funded on a permanent basis. It is a mistake to establish a unit where the personnel are entirely supported by grant funds because this does not allow for the long-term development of unified programs. At the same time, the goal of such a unit must be integrated with the zoo so that institutes do not become so specialized in their endeavors that they have no relevance to the zoo collection. There is a great danger that external funding of research units may result in specialization based on the funding society's needs rather than on those of the zoological institution or society as a whole. One of the primary reasons for establishing a staff answerable to the zoological park itself is to retain control over research goals and priorities, thus preventing the unhappy development of a research unit so autonomous that it could well be geographically located outside the zoo area.

The selection of the core staff should be directed toward individuals who are flexible by temperament and able to carry on a multiplicity of projects within the limitations that any zoological park imposes. The demand for strictly applied research should not be placed on the research unit because nothing will choke off creativity faster. By the same token, the unit should not be so divorced from the collection that alienation is produced, resulting in the lack of free interchange of information.

Ideally, the curatorial staff and the research staff should be integrated so that they are either nearly synonymous or attain levels of communication, good will, and trust, resulting in a research unit which continually feeds into and derives from the collection itself. It may be that the research staff members are not involved at the same levels of participation, but a liaison among some of the research staff members and the animal collection must be maintained. Indeed, the research staff may be instrumental in potentiating the involvement of keepers in data collection as has been outlined by Kleiman (1975). The establishment of communication among the divisions of a park is an extremely subtle point for which there are no absolute guidelines, but it is essential that this be given the utmost consideration in the administrative design of research programs within a zoological park; otherwise, a polarized position will develop where the research unit becomes defensive about its so-called "pure" research activities and develops negative attitudes toward the real demands of the collection.

FUNDING

Funding is a perennial problem for all institutions. In order to establish research activities within a zoological park, the case must be made to the board of trustees or legislature for the need. Presently, there is ample evidence for the validity of research units in zoos, which is documented in such a way that most governing bodies should be responsive and sympathetic (Conway, 1969). It is demonstrable (Zuckerman, 1959) that no major zoological park supports all of its activities from fees paid by visitors. The major zoological parks must be seen as educational components of the community and as cultural institutions. As such, zoos must be accorded financial support by the community where they reside and to which they contribute (Hediger, 1969). As the instructive series of signs for the Brookfield wolf exhibit has shown (Rabb, 1967), research in a zoo feeds back into all activities.

In summary, it is paramount that a core research staff be established that is salaried from a permanent source. This group then serves as a focus for channeling in additional funds and students. Temporary investigators, such as students and postdoctorals, greatly augment the research potential. Given an institutional guarantee of funding for permanent staff positions, external sources of support can be sought with marked success. Even in this time of constricted availability of research funds, there are agencies willing to support small projects if there is a place in which the research workers may carry out their efforts, and, most importantly, if the grant is not to be spent solely on salaries. The World Wildlife Fund, National Geographic Society, Office of Naval Research, and National Institutes of Health have all contributed funds toward projects carried out at the National Zoological Park. Many research grants sought for "pure" research problems have derived much benefit for the collection and have resulted in bringing workers together to create temporary teams that can address themselves to many of the smaller problems of a zoo, which—in and of themselves—would not qualify for separate funding. In short, then, if a permanent research staff exists, the outlook for grant support to carry out projects is excellent.

REFERENCES

Antonius, L. 1938. Uber Herdenbildung und Paarungs eigentumlichkeiten der Einhufer. Z. Tierpsychol. 1:259–289.

Beebe, W. 1952. Introduction to the ecology of the Arima Valley, Trinidad, B.W.I. Zoologica 37(4):157–183.

Bridgwater, D. 1972. Saving the lion marmoset. Wild Animal Propagation Trust, Wheeling W.Va. 223 pp.

Buechner, H. K. 1968. Report from the Smithsonian Office of Ecology: International Program—Research in Ceylon—Smithsonian Year. Smithsonian Institution, Washington, D.C.

Buechner, H. K., H. R. Stroman, and W. A. Xanten. 1975. Reproductive behavior of the sable antelope (*Hippotragus niger*) in captivity. Int. Zoo Yearbk. 14. (In press.)

Conway, W. G. 1969. Zoos: Their changing roles. Science 163:48–52.

Cooper, R. W. 1968. Squirrel monkey taxonomy and supply. Pages 1–30 *in* L. A. Rosenblum and R. Cooper, eds. The squirrel monkey. Academic Press, N.Y.

Crandall, L. 1964. The management of wild mammals in captivity. University of Chicago Press, Chicago. 761 pp.

Eisenberg, J. F., G. M. McKay, and M. R. Jainudeen. 1971. Reproductive behaviour of the Asiatic elephant (*Elephas maximum maximus* L.). Behaviour 38:193–225.

Hediger, H. 1950. Wild animals in captivity. Butterworths, London. 207 pp.

Hediger, H. 1951. Observations sur la psychologie animale dans les parcs nationaux du Congo. Institut Parcs Nationaux du Congo, Brussels.

Hediger, H. 1969. Man and animal in the zoo (Translated by G. Vevers and W. Reade). Seymour Lawrence/Delacorte Press, N.Y. 303 pp.

Kleiman, D. D. 1975. Activity rhythms in the giant panda: An example of the use of checksheets for recording behavior data in zoos. Int. Zoo Yearbk. 14. (In press.)

Mann, W. 1930. Wild animals in and out of the zoo. Smithson. Sci. Ser., Vol. 6. Smithsonian Institution, Washington, D.C. 362 pp.

McKay, G. M. 1973. Behavior and ecology of the Asiatic elephant in southeastern Ceylon. Smithson. Contrib. Zool. No. 125. Smithsonian Institution, Washington, D.C. 113 pp.

Rabb, G. 1967. Educating the zoo public about animal behavior. Pages 14–17 *in* The use of zoos and aquariums in teaching animal behavior. American Association of Zoological Parks and Aquariums, Wheeling, W.Va.

Ratcliffe, H. L. Report of the Penrose Research Laboratory, 1936–1966. Zoological Society of Philadelphia, Philadelphia, Penn.

Ratcliffe, H. L. 1966. Diets for zoological gardens: Aids to conservation and disease control. Int. Zoo Yearbk. 6:4–22.

Schneider, K. M. 1930. Das Flehmen. Zool. Garten (Leipzig) 3:183–198; 4:349–364; 5:200–226, 287–297.

Walker, E. P. 1964. Mammals of the world. Johns Hopkins Press, Baltimore. 1500 pp.

Zook, B., J. F. Eisenberg, and E. McLanahan. In press. Susceptibility to lead poisoning in captive primates. J. Med. Primatol.

Zuckerman, S. 1959. The Zoological Society of London. Nature 183:1082–1084.

DANIEL C. LAUGHLIN

Chicago Zoological Park
Brookfield, Illinois

The Role of Research in Smaller Zoos

To meaningfully discuss the role of research in the smaller zoo at least three separate questions need to be considered: First, is research an appropriate concern of smaller zoos? Second, if so, is one type of research more appropriate than another? And third, what limitations exist for the smaller zoo interested in involving itself in meaningful research and how might some of those limitations be overcome?

IS RESEARCH APPROPRIATE?

If research, in its simplest sense, is the discovery and organization of knowledge (Price and Bass, 1969) and if we can accept the premise that smaller zoos, of necessity, must be actively involved in improving the maintenance and propagation of their exhibited animal species, it then follows that the discovery and organization of knowledge concerning those exhibited animals would be the logical way to improve their maintenance, enhance their propagation, and ultimately ensure their survival. In other words, research should no longer be viewed as a luxury for smaller zoos but as a necessity.

WHAT TYPE OF RESEARCH?

Whether a smaller zoo should concentrate on biomedical or behavioral research, on design or education research, on basic or applied research, on structured or opportunistic research, is dependent on the needs, the resources, the personnel, and the interests of the individual zoo. Such

19

differentiations are perhaps specious, for science is, in its most fundamental sense, an approach to solving problems (Bevan, 1972).

Animal behavior, for example, is so complex that discovery is quite possible. Relatively few species have been studied in any detail in the field, let alone in captivity, and consequently the potential for research is almost unlimited (Rumbaugh, 1971). Furthermore, if understanding animal behavior can improve adaptation and survival in captivity as well as in the field, then the use of zoos for behavioral research is a necessity.

It is a mistake to conclude, as Rumbaugh (1972) has pointed out, that all behavior of captive animals is distorted and worthless as material to understand scientifically. It is true that many kinds of investigations cannot be conducted with animals on exhibit, but there are many studies that can be conducted on questions of social behavior, food preference, response to visitors, infant care and development, activity, communication, locomotion, grooming and play. It seems wasteful to maintain a variety of rare animals primarily, if not solely, for exhibit purposes; more especially when the discovery and organization of knowledge about those animals may assist us in ensuring their ultimate survival.

We have an obligation to utilize the behavioral resources which the animal collections of our zoos provide, and we must utilize these resources in a way that will enhance and enrich the zoo's value and justification for existence.

WHAT LIMITATIONS ON RESEARCH EXIST?

To complete a consideration of the role of research in smaller zoos we must examine the limitations, self-imposed and otherwise, on our ability to conduct that research. We must recognize, as Rumbaugh has stated, that the feasibility of various projects must be determined in light of each zoo's animal collection and its level of access. The determination of individual projects should reflect knowledge on the part of the researcher relating to the intended topic, consideration as to whether the project is feasible in terms of time and zoo resources available, and special interest on the part of the researcher.

It is true that many externally imposed limitations on our ability to conduct research in smaller zoos can not be altered simply by wishing that they no longer exist. However, many limitations can be overcome. For example:

1. One limitation might simply be our attitude toward research in smaller zoos; we can overcome that limitation by changing our attitudes, by recognizing the importance of research to our common objectives.

2. Another limitation might be our own ability; that is, do we possess the knowledge and the skills necessary to conduct research? If we feel that this is a limitation, we can overcome it by self-improvement—by continuing education and by improvement of our observational skills.

3. Another limitation might be a lack of professional assistance—medical, zoological, or otherwise. This limitation may be overcome by the establishment of consultation arrangements with local professionals and institutions. Frequently, many local professionals are eager to assist zoos simply because such work represents a challenging change from their normal routines. Working arrangements with local educational and research institutions can be mutually beneficial and the use of the zoo as a laboratory for a variety of behavioral study programs is quite appropriate.

4. Another limitation might be the priorities we have established within our own institutions. If this is a limitation, we may overcome it by simply re-ordering our priorities with research assigned a greater role in our operations.

One aspect of behavioral research that lends itself to smaller zoos is that all one needs is a pad of paper, a pencil, and keen observational skills. A great deal of supportive equipment is unnecessary.

It should be re-emphasized that research, as simply the discovery and organization of knowledge, is not only an appropriate but a necessary undertaking for smaller zoos. Research is not something that only the larger institutions can or should do. Those of us associated with smaller zoos should not be intimidated, we should not feel as if we are entering a forbidden area and we should not feel that we are intellectually incapable of making research a significant aspect of the operation of our zoos.

REFERENCES

Bevan, Wm. 1972. Research is research is research. Science 176:861.

Price, W. J., and L. W. Bass. 1969. Scientific research and the innovative process. Science 164:802–806.

Rumbaugh, D. M. 1971. Zoos: Valuable adjuncts for the instruction of animal behavior. BioScience 21:806–809.

Rumbaugh, D. M. 1972. Zoos: Valuable adjuncts for instruction and research in primate behavior. BioScience 22:26–29.

L. G. GOODWIN

Zoological Society of London,
Regent's Park

Scientific Work of the Zoological Society of London

HISTORICAL BACKGROUND

The Zoological Society of London was founded in 1826, nearly 150 years ago. Napoleon had been dead for only five years, the Greeks were in the throes of the War of Independence in which Byron died, Wilberforce was fighting the slave trade, and the Spanish empire in South America was at an end. The painter Goya was still alive, Delacroix, Turner, and Constable were at the height of their powers; Walter Scott was writing the Waverly Novels; Keats and Shelley were dead, but Wordsworth was still pouring out poetry. Beethoven's Ninth Symphony had just been performed, Schubert was the favorite of the day, Wagner and Brahms were boys at school. Andrew Jackson ("Old Hickory") was President of the United States of America and the very first 14 miles of railroad had just been opened at Pittsburgh.

In Britain the Prince Regent had at last succeeded to the throne as George IV and he continued to infuriate Parliament with his extravagances. For this we owe him an immense debt of gratitude; his reckless enthusiasm for building came at a good time. John Nash had absorbed and digested the Palladian style and was offered unrivalled opportunities to build exquisitely proportioned terraces, crescents, villas, and churches in London and Brighton. London acquired the Regent's Park, Park Crescent, Portland Place, Regent Street, Buckingham Palace, and a dozen churches. Regent's Park was Nash's masterpiece—the first of the garden suburbs. The old deer park of Marylebone was appropriated to the Crown by Henry VIII when he dissolved the monasteries, and was enclosed in 1540. Queen Elizabeth used to hunt there. After the execution of Charles I the trees were cut down to pay the debts of the Civil War and to provide timber for the Navy; the park became farmland and, during the eighteenth century, grew hay to feed the thousands of horses on which the London traffic depended. Ripe for development at the turn of the century, it escaped the fate of the adjoining fields, perhaps because at one time the Prince Regent had a plan to build a private villa in the park. Instead of being heavily built over like the Portman and Portland estates to the South, a park of about 500 acres was reserved and surrounded with terraces of distinguished houses for the well-to-do. The buildings were erected by private builders relying on private resources; public money was spent only on roads, lodges, open spaces, and railings. The land was leased from the Crown, a terrace facade was built to a design approved by Nash, and the rest of the building was planned in accordance with the builder's own idea. With the exception of Nash himself, the man who put most money and work into the park was James Burton, a wealthy builder. His son Decimus was an architect and designed some of the terraces for Nash; he also planned the layout and buildings of the Zoological Gardens when they were established in 1826 in the northeast corner of the park. Some of his buildings, such as the well-designed giraffe house, are still there.

FOUNDATION OF THE LONDON ZOO

The Zoological Society was founded by a group of scientists and talented amateurs led by Sir Stamford Raffles and Sir Humphry Davy. Davy, president of the Royal Society, was an erratic, many-sided genius; he had invented a safety lamp for miners, had established that chlorine was an element, had purified nitrous oxide so that it could be used as an anaesthetic, and had isolated potassium and sodium by electrochemical methods.

Raffles had spent his energetic, frustrated life working overseas for the East India Company and, in the teeth of opposition from his employers and the intense disapproval of his government, had firmly safeguarded British trade in the Far East by establishing a settlement at the most strategic point—Singapore. He was an indefatigable collector and had seen the animals kept at the Jardin des Plantes in Paris. He was keen to establish a collection of animals in London for the purposes of scientific study and of domesticating new species that might prove to be of economic value.

In a prospectus, probably written by Davy, and issued in 1825, the purpose of the Society was made quite clear:

Rome, at the period of her greatest splendour, brought savage monsters from every quarter of the world then known, to be shown in her amphitheatres, to destroy or be destroyed as spectacles of wonder to her citizens. It would well become Britain to offer another, and a very different series of exhibitions to the population of her metropolis; namely, animals brought from every part of the globe to be applied either to some useful purpose, or as objects of scientific research, not of vulgar admiration.

The objects of the Society, set down in the Charter given by George IV in 1829, were "the advancement of Zoology and Animal Physiology, and the introduction of new and curious subjects of the Animal Kingdom."

It is unlikely that the Founders would have had any idea that, in so short a time, their world—so much of it still to explore and so full of new and curious subjects of the animal kingdom—would shrink so much that the zoo would become a refuge for species in danger of extinction.

In its early days, the Society amassed a great collection of museum specimens—skins, skeletons, stuffed animals and birds—of the greatest scientific importance because many were "type specimens" on which the names and descriptions of the species were based. The Society's museum was eventually amalgamated with the collection of the British Museum (Natural History) in the new building in South Kensington. H.R.H. Albert, the Prince Consort, then president of the Society, chaired the council meeting in 1856 at which the decision to transfer the specimens was made.

The Society has taken its responsibilities seriously. Its monthly scientific meetings have continued almost without interruption for nearly a century and a half; it owns one of the greatest zoological libraries in the world and publishes several scientific journals. It is registered as an "educational charity" and organizes extensive teaching programs for school children and more senior students. It receives no support for its current expenditure from the government or local authority. It maintains a modern veterinary hospital, a pathology laboratory, and well-equipped research laboratories, whose purpose is to study the nutritional requirements, reproductive physiology, and diseases of wild animals and to relate the knowledge gained to problems of man and his domestic animals.

From the very beginning, in 1829, a medical attendant was appointed to care for the health of the animals in the zoo and to report upon them when they died. Close contact was established with the Royal Veterinary College, close by in Camden Town; valuable anatomical studies were made by such famous scientists as T. H. Huxley, William Flower, Roy Lankester, Victor Horsley, John Bland-Sutton, Arthur Keith and Grafton Elliott-Smith.

In 1865 a "Prosector" was appointed to do regular postmortem examinations on all animals. He was James Murie, a pathologist from the

Royal College of Surgeons, widely travelled and very argumentative. He was a source of annoyance to his superiors because the inferences he drew from his examinations were presented as direct criticisms of the management of the Gardens. Too many animals died from injuries, dirt and defective drainage, injudicious feeding, exposure in unsuitable places, absence of sunlight, and general neglect. "That he was correct, there can be little doubt," writes Sir Peter Chalmers Mitchell, who was appointed Secretary of the Society in 1903, "that he was offensive, whether deliberately so or not, no doubt whatever" (Chalmers Mitchell, 1929). He must have done the animals in the collection a valuable service. Murie's successors have included many distinguished scientists, one of them being Lord Zuckerman, the present secretary of the Society.

VETERINARY CARE AND PATHOLOGY

Right from the start, the Society instituted the two scientific services that are essential if captive animals are to be cared for humanely and safely—expert, regular attention to health while they are alive and expert assessment of the cause of death when they die. In no other way is it possible to ensure that animals do not suffer needlessly from disorders that can be put right, or harbor infections that endanger other animals, their keepers, or the public.

The veterinary services of a zoo must have facilities for quarantine and isolation so that newly arrived creatures can be kept apart until they have been thoroughly observed, examined, inoculated, and cleaned up. In this way, the spread of devastating viral infections such as Newcastle disease, and of tuberculosis, parasitic worms, mites, ticks, and fleas can be minimized. In Britain, the Ministry's Rabies Order has enforced quarantine for six months for most imported mammals, and this is all to the good. The most common condition of captive animals that need veterinary attention is accidental injury, and an operating theatre in the zoo is essential if lives are to be saved and deformities corrected. With radiographic equipment, chests can be examined for tuberculosis, and bones for fractures and dietary deficiencies.

The ability to catch animals without undue stress by specially designed facilities built into their quarters, or with the help of tranquilizing drugs in a flying dart, makes proper examination and treatment possible and blood samples can be taken for diagnostic tests.

Laboratory facilities for the culture and identification of pathogenic bacteria and protozoa on the spot make it possible to give warning of dangerous infections so that immediate countermeasures can be taken to safeguard the animals and their attendants.

The veterinary hospital and pathology laboratory of a zoo can generate a great deal of information. Records of the incidence of disease, methods of medical and surgical treatment, differences in sensitivity of various species to anaesthetics, sedatives and antibiotics, special techniques of restraint and management—all these, if made available to others with animals in their care, can lead to general improvements in health and husbandry.

There is a natural disinclination to confess to failure and disaster, whether it arises from misfortune or incompetence, but the information can be of great help to others. It is therefore surprising that so few collections publish records to show how many animals die in the course of the year and from what causes.

All zoos continue to draw heavily on the information collected and published by the Zoological Society of London in countless reports, scientific papers, advisory leaflets, and regular publications such as the *Journal of Zoology, Zoological Record,* the *Symposia,* and the *International Zoo Year Book.* Accounts of the Society's activities are published every two years in the Scientific Reports (1967–1969; 1969–1971).

Our most recent venture has been to collect pathological data, on an international basis, in a form suitable for storage and selection by computer techniques. About a dozen zoos, mostly in Europe, are collaborating in a pilot scheme, supported and aided by the World Health Organization (WHO) and the Food and Agriculture Organization (FAO) of the United Nations; a useful body of information is accumulating.

RESEARCH

For the past ten years the facilities at Regent's Park have included research laboratories (the Nuffield Institute of Comparative Medicine and the Wellcome Institute of Comparative Physiology) with up-to-date equipment for studies in nutrition, reproductive physiology, biochemistry, radiology, and pathology. There is a staff of about twenty research scientists: A few seniors are permanent employees of the Zoological Society and the rest are financed with the aid of project grants from donors such as the Ford Foundation, Wellcome Trust, Nuffield Foundation, Medical Research Council, Overseas Development Administration, and WHO.

We have collected a great deal of information on the hematology of mammals and a monograph has been prepared giving details of the normal cell counts, clotting, and fibrinolytic factors of about 200 species. This is now in the hands of the publisher and will be a valuable reference book for all with animals in their care—when looking for abnormalities, it helps to know that the normal red cell count of a camel is 12 million and of an elephant 3 million, and that the clotting system of the cat family is so active that an equivalent figure found in a man would put him straight into a hospital on anticoagulant therapy. We are studying the nature of the surface membranes of blood platelets, of importance in the formation of clots; thrombosis and embolism are a common cause of death in man but not in other animals. We have also investigated the enzyme (desmokinase) in vampire bat saliva that activates the fibrinolytic system and causes clots to dissolve.

We have made extensive radiographic surveys, partly to detect dietary deficiencies by examining bone density and to collect records for comparative purposes, and partly because of the sheer beauty of the radiographs. They also make excellent teaching material.

Injection of the arteries of dead animals has provided a comparative series on the cerebral circulation, and this will shortly be published as an atlas.

Work on the physiological control of blood flow in cerebral blood vessels has contributed to the understanding of the changes that occur in human subarachnoid hemorrhage.

Nutritional studies have been of help in the design of diets adequate in minerals, vitamins and lipids, particularly for South American monkeys, and comparative studies have thrown some light on the shortcomings of human diets. Much of our food is produced from grassland and contains only small amounts of polyunsaturated fatty acids derived from linolenic acid—essential components of cell membranes, especially in demand during development of the brain.

Investigations are in progress on the immunology of parasitic worm infections, trypanosomiasis, and malaria. The malaria work is carried out in association with WHO, and with other research institutions in Europe and Africa.

Studies of *Mycoplasma* infections have led to new methods of assessing the virulence of strains of the organism, and for measuring the efficacy of vaccines for cattle in Africa.

We have recently been involved in investigations of the cause of death of waterfowl in the London Royal Parks and have shown that these were caused by botulism. *Clostridium botulinum* type C grows in anaerobic mud in warm weather and poisons the birds; the organism can also be transmitted to domestic poultry. We are looking further into this important matter.

Studies of reproductive physiology are clearly of importance in zoos— we look forward to the day when semen from endangered species can be preserved deep-frozen and used for artificial insemination. However, all semen does not survive freezing; human and cattle spermatozoa are very robust but those of pigs and chinchillas, for example, blow up and burst when they freeze. A great deal of basic research is necessary to find the best way of preserving the more fragile spermatozoa, and we have been working at it for the past six years. Last year we sent a research worker to the Kruger National Park in South Africa, where elephants were being culled, to collect semen and freeze it using one of the new techniques he had developed. We now have a stock of frozen sperm that looks good and is highly motile when it is thawed. The next stage—detection of the right time to inseminate the female elephant—is now in hand, and is almost equally difficult. If we succeed, the problems and dangers of breeding elephants in zoos will have been overcome.

We have also given a good deal of attention to the reproductive curiosities of the hystricomorph rodents, relatives of the porcupine and guinea pig. The South American members of this family show some remarkable characteristics, e.g., the tuco-tuco (*Ctenomys*) kept in cages and fed a guinea pig diet, will not breed. The animals develop diabetes and cataracts appear in the eyes. The plains viscacha (*Lagostomus*) ovulates about 800 eggs at a time, implants eight embryos, and brings just two to term.

A study of European bats showed that the females are inseminated before they go into hibernation, but they do not ovulate; the sperm remains quiescent in the vagina all the winter, apparently feeding on the secretions of the mucous membrane. In the spring, ovulation occurs, the sperm springs into activity and fertilization takes place. Young female stoats become sexually mature at the age of about three weeks when they are very small, and still in the nest. They are immediately fertilized by their father (as a sort of insurance policy) but the blastocysts remain free in the uterus until the spring. They then implant and grow to term; like the bats, the young are born at a time of year when the mother can find food for them.

Knowledge of the peculiarities of the reproductive processes of the various species can give guidelines for the successful social grouping and breeding of animals.

FINANCE

All this scientific activity, quite apart from expenditure on the menagerie, costs the Zoological Society of London nearly £½ million a year. Part is retrieved from the sale of publications, but the Society makes a direct annual contribution of about £200,000. Another £150,000 a year is acquired from various organizations in the form of project grants, and this money is, of course, becoming more and more difficult to obtain.

The cuts made by governments in the sums available for research have made competition fiercer and dividends smaller at a time when wages and commodities of all kinds are rocketing in price. And it is not always easy to persuade grant-giving bodies (who are sometimes inclined to be set in their ways and unadventurous in the areas of investigation they support) that there is useful, and perhaps vital, knowledge to be obtained through the study of diseases or physiological and biochemical peculiarities of species of animals they have never heard of. Quite often all chance of support evaporates because no suitable referee can be found to give an opinion on the application.

We have to tell over and over again the truth about our efforts to do original research in zoos. We hold a watching brief for zoologists, physiologists, biochemists, pathologists, for the pharmaceutical industry, and for all biological science. We are on the look-out for things that no one else is looking for. A little is known of the biology of those species that, through the accident of being at a certain place at a certain time, have become "domestic" or "laboratory" animals, or "pests." Practically nothing is known about the rest—hundreds of species of mammals, thousands of birds, reptiles and fish, and hundreds of thousands of invertebrates.

Among them, apart from their innate beauty and fascination, there are "races of Quadrupeds, Birds or Fishes etc. applicable to purposes of utility" and it is through our efforts and our discoveries that others can be led to widen their narrow horizons and to take an interest in them.

REFERENCES

Chalmers Mitchell, P. 1929. Centenary history of the Zoological Society of London.
Scientific report of the Zoological Society of London 1967–69. 1970. J. Zool. (London).
Scientific report of the Zoological Society of London 1969–71. 1972. J. Zool. (London), 166:499–610.

WALTER VAN DEN BERGH

Royal Zoological Society
Antwerp, Belgium

Long-Term Experience of a European Zoo in Research Endeavors

Some of you will remember that when World War II came to an end, the Royal Zoological Society of Antwerp had almost disappeared. Several buildings had been totally destroyed and all the others badly damaged by V-bombs. The zoological collections were almost nonexistent and the botanical collections destroyed; the Society had great debts and the staff had shrunk to some 40 people. There remained only one bright spot: The Society owned the zoo grounds in the center of Antwerp. The site and the buildings had, however, been mortgaged, and in those years we could not count on the help of the public authorities. It was only in 1956 that the first installment of the war damages was paid, the last in 1966.

It is characteristic of man that in times of great trial he finds the energy to rise phoenix-like from the ashes. However hopeless the task before him may seem to be, with the courage of desperation he puts his hand to the plough and succeeds in achieving what seemed impossible to him, because powers about which he had previously been ignorant came to his help. The Royal Zoological Society of Antwerp also experienced this help.

The aim of this paper is not to outline the postwar history of the Society, but it is necessary to be aware of this in order to appreciate our decision as early as 1945 to draw up a program: on the one hand, to determine the policy for reconstructing and organizing the zoo; and, on the other, to lay down a program for expanding our scientific and educational activities.

With regard to the scientific program, we wrote in 1945:

In society, the task of a Zoological Garden is purely cultural and educational. It has to make animals better known from every point of view, to follow them in all aspects of their life and to show their beauty. The animals which have been brought together for this purpose in a Zoological Garden, provide either during their life, or after their death, or sometimes during their life and after their death, extremely valuable material for study for biologists who study the taxonomy, morphology, physiology, pathology, etc., of animals, as well as for those engaged in veterinary medicine. Such study has moreover been of considerable service to the pathological, anatomical, parasitological and other sciences of man himself. This means that a Zoological Garden cannot restrict itself to a simple exhibition of living animals; that all its efforts should help to encourage the study and love of animals, and further that all the necessary facilities should be created in order to allow men of science to make the most use of the material, both living and dead.

To begin completion of the program, the Aquarium was first restored and the Aviary rebuilt as a large number of people in Belgium are lovers of tropical fish and birds.

In 1948, a reporter of *Life* magazine compiled a report on the "Antwerp Cage System" bringing to the Zoological Society considerable international fame.

The Society undertook to carry out all repairs. Well-equipped offices and laboratories were set up to study the zoological collections in a thoroughly scientific way. From the very outset, we were helped by first-rate scientific advisers, who advised us about the organization and equipment of the laboratories. The laboratories were practically completed when *Endeavour* (Vol. 8, No. 29, 1949) published an article by J. Yule Bogue, "Quelques aspects de la conception moderne de laboratoires," satisfying us that, thanks to Prof. Dr. H. Koch, we had not been guilty of any shortcomings.

The then Director of the Prince Leopold Institute of Tropical Medicine, Antwerp, warned us not to be too optimistic for the near future. "The achievements of scientific research," he said, "are rather the result of the man directing the research than of the place where he works; no matter how superb and excellent that place may be, as long as the right man is not there, little progress can be expected." The pertinence of this warning was later revealed.

If we were asked which requirements should scientific research in a zoological garden fulfill, we should answer that since man has the pretension to be at the summit of evolution, he is consequently responsible for the creatures under him, and he cannot allow them to be senselessly killed, or to be deprived of their liberty.

On the other hand, it has generally been accepted that animals may be confined in zoological gardens. If we wish to soothe our conscience, it is clear that it is our duty to try to keep these living beings in the best of conditions; that we should make the most use of this captivity to observe the animals in various aspects of their lives, to study their diseases and to ascertain by all means possible, the exact cause of death. Those organs which are of no use for the autopsy should be placed at the disposal of researchers who know how to make use of them in the first place for the benefit of the animals themselves, and in the second place for the benefit of man.

The skeleton and the skin, insofar as they have not been damaged, should ultimately find their way into a natural history museum. The undamaged parts of the body, not suitable for a natural history museum, should be prepared for the benefit of educational services, as possible illustration material.

Finally, we must bear in mind that one of the principal tasks of a zoological garden is to encourage the animals to reproduce, and scientific research must be concentrated on this.

It is financially impossible for a zoological garden to have its own specialists in all the relevant disciplines, and it is clear that, from sheer necessity, recourse must be had to outside institutes and laboratories having the necessary equipment and specialists at their disposal.

It is also necessary for all scientists who can usefully study the material to know that their collaboration is wanted. This implies that heads of animal laboratories not only have the moral courage to subordinate themselves to the interests of third parties, but also have at their disposal the necessary time and staff to dedicate themselves to a given task.

The head of the laboratory is expected to take care of the animals' health, to give advice about promoting the reproduction of the animals, and most importantly to carry out the autopsies. He will prepare and give the organs which he does not require for ascertaining the cause of death to specialists who ask for them.

It is obvious that the head of the laboratory will have his own scientific program which must serve the problems peculiar to animals in captivity. All data will be centrally recorded on individual cards for each animal: observations during life, treatment given, results of the autopsy, as well as the destination of certain organs and of the skeleton and the skin. This record card will thus give a complete survey of the animal during its captivity and after its death.

The result of such scientific research, including that obtained from outside specialists using material submitted to them, must be published. Between a program as mentioned above and its fulfillment lies a world conditioned by people and money. How has this program been carried out in the Antwerp Zoo?

Although the laboratories were finished in 1948, it was not until 1951 that a suitable head was engaged, the veterinary surgeon, Dr. F.

Andrianne.

Applications had been invited in accordance with the "Regulations for Awarding Research Grants":

1. The Board of Administration of the Royal Zoological Society of Antwerp has decided to award grants for scientific research which is to be carried out in Antwerp Zoo and more especially in the laboratories set up there.

2. These grants are intended

a. to make the material available in the Zoo accessible for scientific research,

b. to improve the living conditions of the animals in captivity and to enable the Royal Zoological Society of Antwerp to profit by the results thus obtained.

3. Research which would cause any pain to the animals cannot be accepted.

4. There will be three types of grants:

a. a grant will be awarded to the holder of a diploma of doctor of veterinary science—the amount will be determined in accordance with the scale of State University salaries and this grant will be awarded for two years,

b. a grant will be awarded to the holder of a diploma of doctor of zoological and psychological sciences—the amount will be determined in accordance with the scale of State University salaries and this grant will be awarded for two years,

c. a grant will be awarded to cover the living and travel expenses of a Belgian scientist who would like to do temporary research in Antwerp Zoo—the amount will be determined separately in each individual case.

5. The candidates must be of Belgian nationality.

6. The application should be accompanied by a well detailed and motivated description of the research project and if applicable by requirements concerning assistance needed.

7. The candidates must supply proof of the fact that they are able to complete successfully the scientific research undertaken.

8. The applications will be examined by a scientific council which will submit its opinions and proposals to the Board of Governors.

9. Only applications approved by this Council will be considered.

10. The applicant undertakes

a. to inform the management of the Zoo of all desirable improvements in accordance with the aim of the grant awarded,

b. the holder of a diploma of doctor of veterinary science undertakes, in addition, (1) to perform the autopsies of animals which die in the Zoo, (2) immediately to make proposals to the Director with regard to the division of the research material available in the Zoo among the various scientific institutions, and (3) to submit after one year to the Board of Governors a report of the progress of his research activity, as well as on the expiration of his grant.

In the period from October 1951 to October 1952, Dr. F. Andrianne served as an excellent collaborator to our program, while at the same time successfully concluding his personal scientific program on "Ornithose parmi les pigeons de Belgique" (Ornithosis among pigeons in Belgium). After a year, however, he left, and it soon appeared that graduates preferred to go to the then Belgian Congo because of the great financial advantages offered to them, and our Society was unable to compete. Not until 1958 were we able to find another scientific collaborator.

The permanent veterinary surgeon continued his work during all these years assisted by the biologist. They performed the autopsies and were assisted by specialists of the Prince Leopold Institute for Tropical Medicine, Antwerp; the Bacteriological Institute for the Hygiene of Domestic Animals of the University of Ghent; the National Institute for Veterinary Research at Uccle, Brussels; the Bunge Institute at Berchem, Antwerp; and the Laboratories of Janssens Pharmaceutica at Beerse.

The record cards of the animal collections were initiated in 1947, with individual cards only for mammals and group cards for birds, reptiles, and fishes. In 1952 the texts of these record cards were harmonized and an individual record card was drawn up for every animal. Later a supplementary card was made for the Laboratory so that all the data after the autopsy could be added to the original record card. This system was soon abandoned and the single record card remained in use. In 1964, a supplementary record card was again introduced to be kept in the building where the animal is housed enabling the keepers, biologists, and attendant veterinary surgeon to add all the necessary observations and data of the treatment at once.

In 1953, the *Bulletin de la Société Royale de Zoologie d'Anvers* began to appear, publishing the results of scientific research in the laboratories. In 1966, the name of this periodical was changed to *Acta Zoologica et Pathologica Antverpiensia.*

In 1958, Prof. Dr. P. G. Janssens was appointed Director of the Prince Leopold Institute for Tropical Medicine in Antwerp. With a clear view to the future, he reorganized the Institute, and late in 1958 proposed that Prof. Dr. J. Mortelmans, a bacteriologist, should be placed in charge of the laboratories. As partial compensation Dr. Mortelmans was allowed to give his lectures on "Bacteriologie et hygiene veterinaire" (Bacteriology and veterinary hygiene) on our premises. This arrangement meant that the interrelation of the Tropical Institute and the Zoo, of which we had been dreaming since 1948, became a reality.

Shortly afterwards the part-time services of the veterinary surgeon, Dr. J. Vercruysse, were secured to look after the animal collections in collaboration with the biologists.

It became clear that a restricted staff could not work ceaselessly. A possible solution to the shortage of staff time lay in collaboration with the planned University of Antwerp. We at once offered our services and promoted the idea of a university having at its disposal valuable zoological and pathological material, much of which still remained unused because of the restricted scientific staff of the zoo. The collaboration which we sought, however, did not come to be since in 1965, when the State University Centre was established, not the slightest notice had been taken of our offer.

Since 1945 we had endeavored to obtain a state grant for our scientific activity. It was only in 1962 that our request was favorably received. In 1967, the one hundred twenty-fifth anniversary of Antwerp Zoo, the grant was trebled. To our great satisfaction and for practical reasons, it was made through a joint committee called the "University–Zoo Association" (Ruca–Zoo). We were convinced that all those who were already associated with our Zoological Society would continue collaboration in the same spirit as before, and that this grant would mean a partial release from the financial burden that the Society had had to bear every year since 1948 in order to build and equip the laboratories and expand its important scientific activity.

We were surprised when the "University–Zoo Association" took the view that attention had to be paid to the letter of the new agreement, rather than to the spirit, which meant that the entire grant had to be spent on new scientific programs. It was taken for granted that the Zoological Society should lose the grant for its scientific activities, which it had previously enjoyed, and should in the future have to bear all laboratory costs alone as well as the programs already in progress at the Zoological Society. Fortunately the Belgian Government intervened, thus relieving us from financial worry.

The possibility of engaging an animal psychologist has never been given up. In 1970 the Zoological Society awarded a grant to a graduate of psychology, in order to enable him to study for the degree of M.Sc. at the University of Edinburgh. After qualifying as M.Sc., however, he was given the opportunity to become an assistant at the University of Ghent. It is anticipated, however, that he will soon be available part-time to our zoo at the expense of the Faculty at Ghent. We hope that this item of the program will be fulfilled in the spirit we had hoped for in our complete scientific program. Further, a collaboration of the "University–Zoo Association" (Ruca–Zoo) is undertaking research into the behavior of fish (*Tilapia*).

Over the years, for educational purposes, we have continually had skulls and limbs prepared of those animals whose skeletons were damaged and could not be considered for one of the two Belgian museums (Royal Belgian Institute for Natural Sciences in Brussels, and the Royal Museum for Central Africa at Tervueren, near Brussels).

We asked Prof. Dr. J. Mortelmans, head of our laboratories and person in charge of our scientific program, to give you a survey of the activities in the laboratories. He submitted the following "Report on the Scientific Activities During the Last Ten Years at the Antwerp Zoo."

The members of the scientific staff of the Zoo of Antwerp were chiefly active in the field of infectious diseases and comparative pathology; the large Zoo collection was profitable, too, for zoological studies in the field of phylogenetics, behavior and anesthesia.

1. Infectious Diseases

Wild animals in captivity are a rich source of parasites and other infectious agents. The particular situation of a dense colony of various animal species kept together in a small place and the continual import provokes a tremendous accumulation of potential epidemics. This unique situation in the field of infectious diseases is an interesting topic for study by veterinarians, physicians, and hygienists. The large mammals and birds and cold-blooded animals of the Zoo constitute a live museum for the public, but the internal and external parasites, bacilli and viruses of these animals do it for the scientists.

Tuberculosis is a major problem in a Zoo. At Antwerp, all the cases of acid fast bacilli were studied bacteriologically and histologically to find out if we have true tuberculosis or infections due to other mycobacteria. Tuberculosis in the Zoo is a veterinary problem, as well as one of public health. Mycobacteria of aquarium fishes may reveal some pathological importance; it became very important when we could infect fish with *Mycobacterium leprae*. Pseudotuberculosis is another prevalent Zoo disease. It is important to study the different types and subtypes of *Yersinia*, since human beings seem to be a serious victim of these germs. Salmonella carriers arrive from all over the world, although Salmonellosis is rare. The question may arise when the bacilli leave the animal body and when this reservoir becomes dangerous for man. The highest Salmonella excretion ratio was found in tortoises.

Wild animals may harbor a lot of viruses communicable to congeners in the Zoo, to domestic animals, and last but not least, to human beings. We had the opportunity to study human hepatitis virus B in chimpanzees and to follow up the dynamics of this infection. The chimpanzee seems to be the only valuable model for the study of this infection. The facilities of the Zoo could provide us with inestimable epidemiological data.

The external and internal parasites which can be found on wild animals at the Zoo constitute a museum of invertebrate parasite animals which is increasing every day. Some of these species may infect human beings; with regard to this latter point, we could make studies in toxoplasmosis, amoebiasis, *Troglodytella abrossarti* and nematode infections in anthropoid apes. Some fundamental parasitological problems, as the dynamics of *Plasmodium falciparum* and *Plasmodium vivax* infections in monkeys, and babesiasis and trypanosomiasis in monkeys, could be studied at our Zoo. The particular localization of some internal parasites offers a unique chance to test out the anthelminthic activity of new drugs. Mycotic infections in apes and monkeys may be compared with human diseases and may serve as a model for the study of pathogenesis and treatment.

2. Comparative Pathology

Chimpanzees were used for studies on arteriosclerosis, lipoproteins and lipid metabolism. This animal species seems to be the ultimate laboratory animal model which can give in this field results useful for human pathology. Long-term dietary trials gave us unexpected results. Baboons and other monkeys could be used, too, for biochemical research in fatty acid metabolism. The aortas and arteries of various animal species were examined to find macroscopical and histological lesions of arteriosclerosis. Spontaneous heart infarct can be studied in mammals and birds which die in our Zoo.

The fine structure of the respiratory mucosa of several mammals was studied by light optical, histochemical and ultrastructural investigation. The findings can be compared with those on the human infant lung.

Several problems of neuropathology and neurobiochemistry encountered in man may be compared with those which can be found or provoked in animals. A lot of naturally occurring neurological diseases could be studied and several surgical interventions were made in monkeys. The blood serum of a lot of animals was analyzed to find a model comparable with human neurological diseases.

The fine structure of the eye of various animals was examined and compared with those of man. This study led us to make some pharmacodynamic trials on the eye pressure in connection with blood pressure, which gave us again some indication on eye diseases in human beings.

3. Phylogenetics

The study of the chromosomes of primates, bovidae and canidae were carried out to give us more information about the phylogeny of these groups of animals. The study of the hemoglobins was most helpful for their zoological classification and history. The interspecies relationships in apes were investigated also by studies of the blood groups of these animals.

4. Behavior

Animal behavior is a fascinating topic in the field, and also in captivity. The stress phenomenon is of some interest for pathologists. But even the study of some problems of environment can give us useful information for the better keeping and breeding of various animal species. Fishes and birds can be easily studied as they do not need large facilities. We could do some work in this field even in apes and other mammals.

5. Anesthesia

The question may arise if anesthesia of wild animals is a tool or a specific scientific branch. It is both. No valuable pathological study or surgical intervention is possible in wild animals without large facilities for anesthesia and even narcosis; for this it is a tool indeed.

But the large range of mammals, birds and cold-blooded animals with the tremendous variation in the biochemical processes of the organism depending on species, age and sex, give us the possibilities of studying reactions of various animal species under controlled pharmacological conditions. The observations made by the application of anesthetic or narcotic drugs can give us indirect information on the variation of the biochemical processes in various animal species. It may give us indirect information about the phylogeny.

6. Conclusion

The collection of wild animals kept by our Zoo gives us the opportunity to make studies in a large range of scientific branches. With regard to the immense possibilities which are available, only a few topics could be explored at Antwerp. Human pathology can find in our Zoo a large field for experimentation. We hope that more and more scientists of our country and from abroad will contribute to exploration of this living museum of vertebrates, invertebrates and microbes.

Scientific research in a zoo is necessary because we cannot, in good conscience, otherwise justify the animals being deprived of their liberty. On the other hand, we cannot imagine that there will ever be a zoo with the necessary financial resources to organize all the disciplines to which modern scientific research has recourse. Therefore, all zoos will remain dependent on the collaboration of third parties—outside scientists.

It is important that public authorities understand that a zoo deserves support because its living possessions are so valuable. However, it is even more important that all scientists who are qualified to serve the zoo with their knowledge help within the framework of their own laboratories.

The "living museum items," which a zoo must keep for generations to come, are mortal, and the time is fast approaching when it will no longer suffice to write out a check in order to replace them in the collection. Even now there are animals which can no longer be bought for gold, and the survival of fauna that will soon be completely endangered largely depends on the intensive study of its way of life, its biotope, its reproduction, and its disease symptoms.

It is of the utmost importance that intensive study be undertaken now for tomorrow it may be too late. Therefore, it is our duty, as zoological gardens, to collaborate as much as possible and to the best of our ability in scientific research, and to encourage everyone who can to help us in our task.

II

BEHAVIORAL STUDIES

ROBERT L. SNYDER

Penrose Research Laboratory, Zoological Society of Philadelphia
and the
Department of Pathology, University of Pennsylvania

Behavioral Stress in Captive Animals

INTRODUCTION

Behavior is defined as the aggregate of observable responses of an organism under given circumstances. Responses or actions depend upon intrinsic psychological, biochemical, and physiological processes. Normal behavior might be considered appropriate responses to various extrinsic or environmental factors which enable an organism to locate food, find shelter from the elements, escape its enemies, reproduce, and rear its offspring. Correct responses have survival value to the organism.

William Cannon (1929), a famous physiologist, and his students (i.e., Cannon and Rapport, 1921) demonstrated that mammals, including man, react with increased activity of the autonomic nervous and hypophyseal–adrenal hormonal systems, when facing physical danger. This acute stress reaction includes increased secretion of adrenaline, a definite shortening of the clotting time of the blood, a rise in metabolic rate, tachycardia, increased cardiac output and high blood pressure. All of these mechanisms Cannon described as purposeful ones, providing to the mammal optimal conditions for fighting or flight. He further suggested that chronic functional disturbances may develop when these reactions occur in inappropriate situations or cannot be extinguished by the release of physical activity. Naturally, many research biologists consequently postulated that both the acute stress reactions and those more prolonged might contribute significantly to the development and course of certain diseases.

Supported by grants from the National Heart and Lung Institute (HL-1979) and the Smith Kline and French Foundation.

Hans Selye (1959) defined stress as the common features in the reactions of living organisms to all stimuli that tend to disturb the dynamic equilibrium of psychological, biochemical, and physiological processes. If prolonged and intense, the "stressors" induce a sequence of reactions which Selye calls the "General Adaptation Syndrome." The endocrine or hormonal adjustments that occur during stress are emphasized particularly as being useful to the organism in maintaining homeostasis. He proposed the hypothesis that the adaptive mechanisms called into operation may derail and cause disease or interfere with reproduction if the noxious stimuli are prolonged and unrelieved. For example, the secretion of large amounts of corticotrophin (ACTH) and anti-inflammatory adrenal steroids may be useful in enabling an organism to survive during an emergency by suppressing excessive inflammatory reactions; but, on the other hand, the same response may be harmful inasmuch as it permits the spread of infectious agents.

The most essential and probably most creditable aspect of Selye's concept of stress is that mammals respond in predictable ways to such diverse stimuli as hunger, infection, burns, toxins, trauma, muscular exertion, rage, and fear. The most publicized adaptive responses are increased adrenocortical function, involution of the thymicolymphatic system, and deceased reproductive function. The exact mechanisms involved are too complex to cover satisfactorily in this review. However, the general concept is simple enough. The basic endocrine and physiological adaptive mechanisms serve to maintain internal physiological homeostasis in the face of noxious external stimuli and to equip the animal to meet the demands of emergency situations. The desirable goal of internal stability is not achieved without sacrificing functions less immediately vital to the individual, such as growth, reproduction, and resistance to infectious disease and parasitism.

Christian (1959a) subsequently discovered that sociopsychological interactions activate these adaptive responses in mice, moles, and Norway rats. Increased adrenocortical activity is related to social dominance–subordinate relationships with the more subordinate animals exhibiting a greater response than the dominant ones. The intensity of social interaction is related to population density because opportunities for contact increase with numbers (Davis, 1951a, b; Christian, 1955a, b, 1956, and 1959b, c; Christian and Davis, 1955, 1956; Christian and LeMunyan, 1958; Christian and Williamson, 1958; Snyder, 1961, 1967, 1968; Ratcliffe, 1968).

Unfortunately, the emphasis on population density might have obscured the real meaning of Christian's work. It may also explain the obsession of some zoo critics with crowding as a principal fault in zoological gardens. The primary stimuli are still the psychological responses of the animal to various features of its environment. Stress depends upon how the animal in question perceives whatever is happening around it. What is alarming to one animal may be gratification for another. Thus, behavioral stress is not merely a matter of how many animals in a cage, but what kind of animals, age composition, sex ratio, complexity of the cage environment, prior experience of the occupants, and so forth.

That behavioral stress affects growth, reproduction, and resistance to infectious and parasitic disease in caged mammals is reason enough to consider whether the principle of adaptive responses can be utilized in zoo management. Zoological gardens maintain many different species of vertebrate animals in widely varying environments which are very different from their native habitats. A number of environmental factors inherent in zoo management can easily be interpreted as having stressor potentialities. Caged wild animals are faced with a number of problems not encountered in their natural state. Potential stressors might be listed as follows, not necessarily arranged according to their intensity: restricted movement, lack of concealment, unfamiliar food, interrupted circadian rhythms, abnormal social situations, sudden environmental changes, crowding, and a sterile environment (the well-known naked cage).

Abnormal social situations include those in which males and females are unable to disperse after the breeding season or where the young are unable to disperse after maturity. These would also include circumstances where animals were isolated or where young were reared without juvenile companions or in the absence of the appropriate parent or parents. Sudden environmental changes include nearby unusual activities, noises, switching cages, transportation to other zoos, introduction of new cagemates and loss of accustomed cagemates. All of these factors are potentially capable of producing either acute or chronic stress.

Species normally exhibited in zoos belong to four classes of vertebrates: Amphibia, Reptilia, Aves, and Mammalia. Generally, study of the adaptive responses has concentrated on mammals but it should be mentioned that amphibians, reptiles, and birds have similar hormonal and autonomic responses. Investigations have demonstrated that birds react to social stress with increased adrenocortical activity and depressed reproductive function (Siegel, 1959; Flickinger, 1961, 1966).

The express purpose of this report is to focus attention on research that has demonstrated how environmental factors can induce behavioral stress in captive wild animals. Theoretical aspects of animal behavior are discussed briefly to guide the uninitiated to the literature. A number of techniques in zoo management are described because the usual laboratory techniques for handling and restraining experimental animals may not be appropriate to captive wild animals. Also, handling and restraint—either for study purposes or for routine management—elicit stress nearly as intense in some instances as do other recognized environmental stressors. Failure to recognize such procedures as stressors might easily confound experimental results. Parts of the following sections are devoted to what is known already about the behavior of zoo animals and the effects of the zoo environment on behavior. Conditions vary from one zoo to another and scientists and investigators should know how these variations might affect the results of their studies. Finally, the literature on diseases of zoo animals is reviewed to illustrate the intricate reciprocal relationships of disease and behavior.

THEORETICAL ASPECTS OF ANIMAL BEHAVIOR

Important theoretical considerations in the study of animal behavior are particularly pertinent to work in the zoo. The trained ethologist, already well informed on the subject, can ignore the following discussion meant only to guide the scientist whose expertise is in another field.

Biologists generally agree that animals possess predictable behavioral patterns; hence, all behavioral sciences are based on the assumption that predictions about behavior can be made if a sufficient number of relevant variables are known. How these patterns of behavior are acquired is still a matter of dispute.

Theorists generally agree that behavior always has a cause. Stated in another way: There must be a reason why an animal behaves the way it does. The vitalists postulated as final causes entelechial soul-like factors and unfailing inexplicable instincts. One school of psychologists in America emphasizes that behavior is purposefully directed to a specific goal. In this context, an animal is motivated by expectancies. The mechanistic schools, which include reflexologists such as Pavlov, are convinced that all behavior can be explained in the final analysis by physical laws. Another recognized school of thought, particularly strong in America, is termed behaviorism. J. B. Watson (1919), the credited founder of behaviorism, stressed the influence of the environment on behavior and was especially concerned with its effect on learning. Extreme proponents of these viewpoints maintain that all behavior is learned (acquired).

Phylogenetically determined behavior is termed innate or instinctive and develops in the absence of possibilities for learning or imitaton (see Eibl-Eibesfeldt, 1970, and Lorenz, 1970). Innate behavior patterns need not be functional in the neonate but develop later in life under certain circumstances or in response to specific stimuli. Phylogenetically determined behavior is adaptive and has definite survival value.

The basic arguments today among students of behavior are concerned with which behavior patterns are innate and which are learned during the course of ontogenetic development. Ethologists design experiments wherein the animal is isolated from other conspecifics, thus preventing any imitation and making it impossible to learn the behavior in question by trial and error. Critics of this experimental design (Lehrman, 1953, and Hinde, 1966) argue that it is impossible to raise an animal without any experience, because it is always a part of an environment, even within the egg or the uterus, and that it is always experiencing something while interacting with its environment. A paper by Kuo (1932) is frequently cited as an example of prenatal acquisition of a behavior pattern. In response to Kuo's observations, Lorenz (1961) called attention to other species of birds having similar experiences within the egg, but which gape, strain the mud for food, or put their bill into the mouth of their parents.

The argument about innate and acquired behavior patterns and the effects of the environment on learning is especially applicable to zoo research, because different animals of the same species living in dissimilar zoo environments naturally have different opportunities to learn. Innate behavior that is frustrated in the zoo habitat may lead to stereotyped or abnormal behavior patterns. Furthermore, appropriate behavior in the natural environment may be inappropriate in certain zoo habitats.

As curators do not generally include ethograms (an inventory of behavior patterns) among their records, it is impossible to determine an animal's previous experience; thus, one is not able to determine why a particular zoo animal behaves the way it does or what part of its behavioral repertoire is innate or otherwise. Unfortunately, there are few books devoted exclusively to the behavior of captive wild animals. Heini Hediger, director of the Zurich Zoo, has published several papers on this subject in German from 1938 to 1970. Two of his books have been translated into English (Hediger, 1950, 1969). Lee Crandall's book published in 1964 is a comprehensive reference to techniques of managing wild mammals in captivity. Herbert Fox's book (1923) is basically a description of patterns of diseases in wild mammals and birds in the Philadelphia Zoo but does mention problems related to abnormal behavior of wild animals in captivity.

An excellent reference source for the zoo research biologist is the 10-volume International Zoo Yearbook published by the Zoological Society of London. Each volume contains a collection of papers covering a special topic: 1—apes in captivity; 2—elephants, hippopotamuses, and rhinoceroses in captivity; 3—small mammals in captivity; 4—aquatic exhibits in zoos and aquariums; 5—ungulates in captivity; 6—nutrition of animals in captivity; 7—penguins in captivity; 8—canids and felids in captivity; 9—amphibians and reptiles in captivity; and 10—birds of prey in captivity.

TECHNIQUES IN ZOO RESEARCH

Characterization of the Experimental Animal

Individual identification of each animal is a prerequisite to reliable record keeping. The larger mammals and reptiles may be individually identified by cage assignment when there is no danger of mistaken identity. In some cases, identifying scars and physical characters are reliable. The widespread use of tattooing and numbered metal ear tags for mammals and leg and wing bands for birds is fortunately employed in most of the larger zoos.

Most zoos begin a record when the animal is acquired and added to the collection. Such records should include geographic origin, method of capture, age, weight, physical measurements, length of transit, type of transportation, quarantine periods, and other pertinent information. Records in the zoo should include cage assignments, identity of cagemates, breeding performance, vaccinations, immunizations, illnesses, drugs administered, surgery, and a profile of behavior.

Identification of the Animal

It is axiomatic that the animal studied should be correctly identified as to species. Unfortunately, this is not always easily accomplished, especially when a specimen cannot be furnished to a museum. The investigator should first consult the curator in charge of the animal in question as this official in the zoo is responsible for the identification of animals in the collection. Inasmuch as the animals in the zoo are alive and cannot always be restrained conveniently for measurement of taxonomic characters, the classifications used are sometimes tentative. Correct identifications are ultimately established when the animals are examined after death, but longevities in some cases are such that it is impractical to wait for confirmation of tentative identification. As an alternative to museum identification for studies involving behavior, color photographs of typical specimens should be provided for publication. The photograph will not provide exact criteria for classification in all instances, but it is better than nothing.

The investigator should always keep in mind that mistakes in classification are possible. A notable example involves the largest of the anthropoid apes, the gorilla. It is now generally agreed that there are two recognizable races presently separated by several hundred miles of jungle area: the lowland gorilla (*Gorilla gorilla gorilla*) and the mountain gorilla (*Gorilla gorilla beringei*) (Allen, 1939). The most obvious characters distinguishing the two subspecies are the shorter arms, the longer black hair, and the narrower, more lengthened face of the mountain gorilla (Coolidge, 1929;

Schultz, 1934). Also, the adult mountain gorilla is said to attain a greater weight (Willoughby, 1950). The mountain gorilla brings a higher price in the zoo market, but the lowland gorillas are easier to come by; thus, animal dealers prefer to classify their merchandise as mountain gorillas regardless of their geographic origin. Most gorillas are only infants when sold to the zoo, in which case the curator—with only comparative features (shorter than, longer than, narrower, etc.) as guidelines—is hard pressed to make a correct identification.

Taxonomists presently disagree on the necessity for separating species into races. Dobzhansky and Epling (1944) define races as "populations of a species which differ in incidence of one or more variable genes or chromosome structures." The variable genes or chromosome structures may not control a morphological character, however, in which case the taxonomist would not recognize the race. Physiological differences, as well as morphological distinctions, between spatially separated populations are to be expected (Bailey, 1939, DuShane and Hutchinson, 1944). There are also several examples in the literature of populations of the same species which are morphologically identical but which have different behavioral characteristics. Ricker (1938) studied a population of the sockeye salmon (*Oncorhynchus nerka*) in Cultus Lake, British Columbia, which were divided into residual and migrating components. The residual population is the progeny of the migrating population, and no reproductive isolation or speciation is indicated. Another pertinent example concerns the three kinds of crickets (*Nemobius fasciatus fasciatus, N. f. socius,* and *N. f. tinnulus*) studied by Fulton (1933), which are strikingly similar in morphological characters. The chirps of the three forms are distinct and, under natural conditions, probably prevent the populations from interbreeding. Fulton was able to raise fertile hybrids of *fasciatus* and *tinnulus,* proving there was no genetic isolation.

These examples serve to point out that animals in the zoo that are morphologically similar to one another may have different behavioral characteristics by virtue of their racial or subspecific peculiarities. Such differences could easily lead to misinterpretations of results; for example, in comparing the behavior of a species in two different zoological gardens. Errors of this type can best be eliminated by noting the geographic origin of zoo specimens whenever possible. In this respect, the importance of accurate records of acquisition for zoo animals cannot be overemphasized.

Handling and Restraint

Wild animals in general are more excitable than either domestic or laboratory animals and infinitely more difficult to handle and restrain. Zoo managers, having had so much personal experience with animals severely injuring or killing themselves because of an unusual disturbance, are reluctant to allow research workers unrestricted access to the animal quarters. At first, one might find it hard to believe that one more human in the animal's environment would be noticed, since thousands of zoo patrons pass by the exhibition cages every day and animal keepers are cleaning cages and feeding animals from roughly 8:00 a.m. to 5:00 p.m. each day. Apparently, reactions depend on what the animals become accustomed to. Customers that stay on the walks behind the barrier fences are invariably ignored during the normal zoo hours. But the same patrons crossing the barrier fences or coming into the zoo before and after the usual public hours are cause for alarm. An appropriate generalization in managing captive wild animals is to avoid sudden changes in the environment.

Excited animals are liable to injure themselves running into the bars and fences, but this is not the only consequence of alarming stimuli. Christian and Ratcliffe (1952) documented 14 examples of what they termed fatal shock disease in the Philadelphia Zoo. The animals, carnivores of small to medium size, relatively young and well nourished, all died after being subjected to relatively minor stress, such as being transferred to new quarters or being disturbed by repairs to adjacent cages. Survival time after the distressing stimuli varied from a few hours to two days. These investigators suggested that the immediate cause of death was hypoglycemia, which reflected their inability to respond normally to sustained stimulation because their adrenal cortices and, in some instances, their pituitaries had undergone partial atrophy through inactivity. These animals had all been caged in small indoor cages.

Zoo managers naturally wish to limit handling and restraint of their animal charges. Research workers will find that most zoos have formulated general policies which rule out experimentation involved directly with the animals. In practice, experimental surgery, drug trials, collection of body fluids and tissues, feeding experiments, and, in some cases, manipulation of social groups for behavioral studies are all strictly prohibited. Such policies, however, are not detrimental to the majority of research projects designed for captive animals, for example, studies based on collection of blood samples. The policies described above would not allow investigators to capture and restrain animals specifically for the collection of blood; however, permission would be granted if the blood were collected when the animals were being captured and restrained for other purposes, such as trimming hooves, removing antlers, cutting claws and toenails, cleaning teeth, or treatment for illness.

To prevent undue suffering, euthanasia is required for injured and sick animals that are beyond treatment. At this time, just prior to death, the required materials can be collected without circumventing the rules against direct intervention. Injury, unfortunately, is still the leading cause of death in most zoological gardens, and many of the victims are healthy specimens that can fill the requirement of many studies for normal, baseline measurements.

Many research programs only require access to the dead animal to collect the required material. In this case, the investigator should make arrangements for dead animals to be refrigerated immediately after death. A tremendous amount of material can be salvaged from thorough postmortem examinations.

Mechanical Restraint A squeeze cage has one movable side which can be moved to press an enclosed animal gently but firmly against the bars at the opposite side. These restraint boxes are usually portable and can be moved to the animal's home cage and bolted to the door frame. The use of such devices involves various levels of stress which depend upon species characteristics, individual characteristics, age, prior experience, and duration of restraint. Only a few procedures such as tattooing, fastening ear tags, giving injections, applying topical medicines, and trimming claws and toenails can be accomplished satisfactorily on an unanesthetized animal in this device.

Some restraining devices are designed as an integral part of the animal's home cage. Cage layouts in most modern zoos allow for shifting animals to adjacent cages to allow the keepers access for cleaning and repair. A squeeze box designed as a passageway between cages has many advantages over the portable type. Since resistance to stress is related to prior experience (Christian and Ratcliffe, 1952; Weininger, 1954; Bernstein and Erlick, 1957; Hatch *et al.,* 1963; Ader, 1965, 1966, 1969; Ader and Friedman, 1965; Barrett and Stockham, 1965; Goldman, 1965; Friedman and Ader, 1967), it is desirable to provide the animals with experience in the squeeze box before manual handling is practiced. A few minutes of restraint at periodical intervals reduces the risk to the animal and makes physiological measurements more meaningful.

Nets are frequently used in the zoo to capture, restrain or move animals from one cage to another. All techniques used involve considerable physical and psychological trauma but their use is unavoidable under certain circumstances. The research investigator should be aware of the effects of handling on various physiological, biochemical, and behavioral parameters in laboratory animals and should assume that the effects will always be greater.

Immobilizing Drugs and Tranquilizers   A number of drugs have been used experimentally to facilitate handling wild captive animals. The investigator is reminded that the same drug produces different effects in different species and also that anesthesia, immobilization, and tranquilization are not one and the same. An animal can be immobilized and still feel pain and it can be tranquilized and still move. Furthermore, immobilized, anesthetized, or tranquilized animals may still be subject to stress. Also, each drug has its own peculiar effect on such important parameters as blood pressure, heart rate, and adrenal steroid and catecholamine release.

Sometimes, these drugs can be concealed in food without arousing the animal's suspicion. On the other hand, an animal, once immobilized in this fashion, often may not be duped a second time. An aged male gorilla in the Philadelphia Zoo, successfully immobilized once with a drug in his daily cup of orange juice, refused fruit juice for six years afterwards.

Sodium pentobarbital is an example of a drug which can be given orally to produce general anesthesia in the large carnivores. The approximate dose is 10 mg/lb body weight, a little less in obese or aged animals. In most instances, body weights can be furnished by the curator.

A useful instrument for administering drugs to wild animals is the air gun, which propels a hypodermic needle and loaded syringe accurately up to 40 yards. The propellent is carbon dioxide delivered from a pressurized metal cylinder. The plunger in the syringe is driven forward at impact by a small explosive charge. Injections can be delivered subcutaneously or intramuscularly by selecting the proper length needle with or without a collar.

Complete immobilization can be accomplished with drugs that either paralyze or anesthetize. Tranquilizing drugs are sometimes used to partially immobilize an animal or render it tractable. The selection of the appropriate drug depends on the species of animal, duration of restraint, and degree of restraint required.

The principal advantages of the paralyzing drugs are their speed of action, clear separation between normality and paralysis, speed of recovery, and simplicity in use.

Paralyzing drugs have certain disadvantages too. Most important perhaps is the fact that with the possible exception of the nicotine alkaloids, which are no longer recommended, the paralyzed animal is fully conscious. The amount of distress caused the animal in this way may have a marked effect on biological data collected. Heart rate, blood pressure, and adrenal cortical steroid blood levels, for example, are elevated quickly after an animal is alarmed. It should also be understood that paralyzing drugs may lack analgesic properties.

The literature on immobilizing drugs is widely scattered and not very extensive. It would be impossible to cover many species in this limited report, but a few general statements and a list of pertinent papers on the subject might be helpful.

Gallamine triethiodide (Hall *et al.,* 1953) and the nicotine alkaloids (Crockford *et al.,* 1957) were probably the first paralyzing drugs used to immobilize wild animals. Gallamine triethiodide is a synthetic substance with curariform properties marketed under the trade name of Flexedil. Prostigmin is an antagonist to the neuromuscular block action of this drug. Some pitfalls to be avoided in using Prostigmin are discussed by Harthoorn (1965). Nicotine has many undesirable side effects and its use is no longer recommended.

Succinylcholine is an effective agent to immobilize animals for a short period of time. It possesses certain advantages that account for its past popularity. Even if injected intramurally, it has a rapid onset, and even a very small amount is effective. The recovery period is likewise short and the animal usually recovers to its feet with minimal struggle, which reduces the danger of self-inflicted damage.

Succinylcholine has several disadvantages. As there is no antidote for the drug, overdoses are invariably fatal, and the necessary dosage varies tremendously with different species. The amount required is proportional to the amount of cholinesterase present. This may be gauged by the amount of cholinesterase or pseudocholinesterase in a sample of blood plasma (Harthoorn, 1965). The injection of succinylcholine causes a rise in blood pressure and also a rise in serum potassium (Stowe, 1955; Stevenson and Hall, 1959; Hofmeyr, 1960; Tavernor, 1960; and Schleiter and Casper, 1961). Human volunteers describe extremely painful muscle cramps associated with the paralysis. All of these effects combined with full consciousness would constitute an extremely stressful experience to an experimental animal.

Succinylcholine seems particularly suited to the smaller African antelopes and members of the family Cervidae. For example, dosage levels for white-tailed deer (*Odocoileus virginianus*) range from 0.033 to 0.047 mg/lb of body weight (Allen, 1970). Hyaluronidase injected with the drug reduces the latent period (time from injection to paralysis) from 5.1 to 2.3 minutes and allows a 15-percent decrease in dosages (Allen, 1970). The net result of this absorption-promoting enzyme should be to reduce toxicity and, hence, mortality.

With the exception of the hippopotamus (Harthoorn, 1965), succinylcholine is not recommended for the larger hoofed mammals. An oripavine derivative with potent morphine-like activity, identified almost everywhere in the world as "M.99," is an effective immobilizer for the giant mammals in the zoo. The effects of this substance can be reversed with the chemical nalorphine. This drug is especially appropriate for giraffes, rhinoceroses, elephants, and zebras because the dosage can be meaured to allow for the animal to remain standing. For the long-legged giraffe and the ponderous elephant, this condition reduces mortality from lung hypostasis and decreases the risk of injury.

The investigator using M.99 either alone or in combination with other drugs should be aware of their effects on biological parameters. Sight is impaired by both M.99 and hyoscine, and visual accommodation is lost (Harthoorn and Bligh, 1965). Hearing remains at normal acuity and alarming sounds result in immediate flight reaction. Under M.99 influence, apprehension and fear are banished but curiosity seems to be unrestricted.

Few animals went down at the dosages of M.99 employed by Harthoorn and Bligh (1965). With ruminants, this is particularly important because they may otherwise regurgitate ruminal contents or become tympanitic. Ruminal movements are in complete abeyance under the influence of M.99, but expulsion of gas by eructation is possible.

The recumbent animal will usually rise within a minute or two of receiving an intravenous injection of nalorphine, which is a nonspecific antidote to M.99. The dose of nalorphine is based on the size of the animal rather than on the dose of M.99. The dose of nalorphine for a 2,000-kilogram beast is 250–500 milligrams as compared to 40–60 milligrams for a zebra (Harthoorn and Bligh, 1965).

Harthoorn and Bligh (1965) found that the subject does not return quickly to complete normality after the nalorphine and, therefore, should not be exposed to cagemates for at least 6 hours. That nalorphine is a respiratory depressant was also noted.

M.99 raises the blood pressure on conscious sheep and the heart rate in conscious sheep and donkeys. Harthoorn and Bligh (1965) found that these changes also appear to occur in wild ungulates, so that all stressors should be avoided while the animal is under the influence of M.99. Animals to be transported should be given small doses of nalorphine followed immediately by a tranquilizer such as chlorpromazine.

Phencyclidine   The application of phencyclidine [1-(1-phencyclohexyl) piperidine hydrochloride] as an immobilizing agent has until recently been primarily directed at the primates (Kroll, 1963; Domino, 1964; Melby and Baker, 1965; Vondruska, 1965; Krushak and Hartwell, 1968). This drug acts principally on the central nervous system (Domino, 1964). With certain species and under certain circumstances, it may have advantages over paralyzing drugs. The narcotized animal shows a minimal fear and has at least some analgesic resistance to pain. Death from overdosage of narcotics is less likely than from paralyzing drugs due to the larger safety margin.

Injected alone, its effects vary according to the species. Generally, it causes a cataleptic state in low to medium dosages and a condition resembling anesthesia at higher dosages. The animal receiving a low dosage appears to be wide awake but unable to move; the pupils are dilated and respond sluggishly to changes in light intensity. There is little relaxation of muscles, and hypertonicity, especially of the forelimbs and neck, usually is evident. Trembling and light clonic or tonic spasm involving all the body musculature may be present, particularly in dogs (Kroll, 1963; Domino, 1964; Zaldivar, Jose and Otter, 1967). Phencyclidine injected 1 mg/lb of body weight invariably induces a rhythmic head weaving in woodchucks (*Marmota monax*) (Snyder, unpublished data). Similar uncontrolled weaving head movements have been reported for wolves, coyotes, red foxes, and leopards during the recovery stage.

Since the swallowing reflex is generally not affected, medication can be given orally to immobilized primate species when no other method is possible (Snyder, unpublished data). On one occasion, an acutely ill siamang gibbon (*Symphalangus syndactylus*), which had stopped eating, was given antihelminthetic and antibiotic drugs while immobilized with phencyclidine. The drug was mixed with milk and introduced into the mouth by cup. The animal swallowed 50 milliliters of liquid without aspirating any portion of it. Such a procedure is extremely valuable to know because captive animals are overly suspicious of additives even when they are eating well. Excessive salivation is a common response to this drug, but it can be counteracted with atropine (0.5 mg/0.10 lb).

Hallucination has been reported in human subjects (Levy, Cameron, and Aitken, 1960) and apparently this is a potential side effect for cats (Harthoorn, 1962) and gorillas (Snyder, unpublished data).

Animals recovering from prostration by phencyclidine retain the hyperextension of the forelimbs and stiffness of the neck muscles well into the recovery stage. Animals make repeated effects to stand only to fall because of lack of coordination. Depending on the dosage used, the recovery stage may last from 2 to 6 hours.

Kroll (1963) and Seal, Erickson, and Mayo (1970) described the results of experiments using phencyclidine alone on many mammalian species in the zoo. Undesirable effects of this drug were occasional excitement intensified by visual and auditory stimuli, skeletal muscle hypertonicity in the recumbent animal, frequently increased salivation, nystagmus, fasciculations and a 1–5 °F rise in body temperature in many species. No instances of failure of immobilization were encountered with doses from 0.5 mg/kg to 2.0 mg/kg. The time required for complete immobilization varied from 10 to 40 minutes, while recovery stages lasted from 1.5 to 3 hours with effects noted for 6 to 18 hours, including decreased activity and depressed responsiveness to external stimulation.

Seal, Erickson, and Mayo (1970) subsequently proved that promazine greatly reduced the side effects associated with the use of phencyclidine alone. The replacement of muscle hypertonicity with relaxation, the control of the hyperthermia and salivation, and the virtual elimination of convulsive activity were of particular value. Thus far, no serious consequences to the fetus or to pregnant females have been observed.

This same research team has used a combination of promazine and phencyclidine in 2,005 trials involving 1,075 animals of 156 species of mammals in zoo situations with only two deaths. Their results suggest that phencyclidine should be considered the drug of choice for immobilizing primates and the Carnivora in the zoo. Convulsions occurred in some animals, but these were controlled easily by the administration of additional promazine. Fewer trials involved species of Rodentia, Tubulidentata, Pholidota, Insectivora, Hyracoidea, Marsupialia, and Edentata, but even so the drug combination would seem to have advantages over other immobilizing drugs for members of these orders. Investigators contemplating the use of phencyclidine should refer to the literature for dosage schedules.

Phencyclidine is less efficacious for immobilizing ungulates or the larger animals because of the long recovery stage. These animals may cause considerable damage to themselves during this period by stumbling or running against the fences and walls of their enclosure. Certain animals in the zoo can be restrained manually during this period but still the time required is costly in terms of tying up the animal keepers.

What applies to the ungulates, however, applies also to the primates and Carnivora. One should remember to allow the immobilized animal to recover in a more confining chamber or even a shipping crate to prevent the animal from injuring itself. For example, monkeys and apes must be confined so as to prevent them from climbing during the recovery stage. The same precautions are also required for the initial immobilization stage, for a frightened primate will usually climb to the cage ceiling and may fall when the drug takes effect.

Experiments have been conducted in which tranquilizers were used at relatively high dosages to immobilize zoo animals. In general, these drugs, even at high dosage levels, are not satisfactory substitutes for the narcotic and paralyzing drugs, as the animals are never completely immobilized. At best, an animal is drugged sufficiently to render it more tractable to mechanical or manual restraint. Tranquilizers are recommended principally to counteract the side effects of the immobilizing drugs or to reduce the level of stress when animals are captured, handled or transported.

Major tranquilizers, or neuroleptics such as reserpine and chlorpromazine, have been shown to reduce fighting behavior in mice (Christian, 1957, and Vessey, 1967). Diazepam, one of the benzodiazepine compounds, is an antianxiety agent which produces muscle relaxation in conjunction with its sedative effects. It is classified as a minor tranquilizer. Diazepam has been reported effective in reducing aggression in a wide variety of species and circumstances (Randal, *et al.,* 1961; Valzelli, Giacalone, and Garattini, 1967).

Various tranquilizers are routinely employed in the zoo to reduce the level of aggression in animals caged in groups. However, the assumption of reduced aggression in these situations is not always warranted. Diazepam applied at low doses, for example, appears to augment rather than depress agonistic behavior in groups of male house mice (Fox and Snyder, 1969).

Studies of Olds and Baldrighi (1968) suggest that low doses of diazepam or chlordiazepoxide, closely related antianxiety agents, increase sensory inflow to the rat brain. Similarly, Schallek and Kuehn (1965) noted behavioral restlessness and increased EEG frequency in cats given low doses of these drugs. It is well known that, in the mouse, olfaction plays a major role in determining the nature of the animal's interaction with its social and physical environment (Ropartz, 1968).

Tranquilizers are not without their side effects. Thus, continued use of these drugs to alter social behavior in caged wild animals should be considered as an experimental variable. Reserpine and chlorpromazine, while exerting a tranquilizing effect on the central nervous system, have been reported to inhibit the release of gonadotrophins, especially luteinizing hormone in the rat (Gaunt et al., 1954; Zarrow and Brown-Grant, 1964). In addition, these substances cause augmented release of prolactin (Talwalker et al., 1960; Meites, Nicoll, and Talwalker, 1963) and of ACTH (Gold and Ganong, 1967). While this increase of ACTH secretion was persistent in some studies, in others reserpine or chlorpromazine appeared to block the increase of ACTH accompanying another stressor (Gold and Ganong, 1967). Giuliani, Motta, and Martini (1966) have proposed an explanation for this apparent contradiction. The main effect of reserpine on the pituitary–adrenal axis is to enhance ACTH secretion, but on the other hand the blocking of stress reactions could be due to the feedback effects of augmented adrenal steroid secretion produced by the reserpine.

Chlorpromazine has been shown to delay implantation in the rat (Psychoyos, 1963). Delay of implantation occurs because of interference with the estrogen–progesterone sequence of secretion (Psychoyos, 1966), and tranquilizers are thought to produce their effect by inhibiting gonado-trophin release and thereby blocking secretion of ovarian estrogen (Psy-choyos, 1963). Cortisone can prevent some of the effects of reserpine and chlorpromazine on inhibition of gonadotrophin secretion (Chatterjee, 1965, 1966).

Reserpine or chlorpromazine treatment from day 6 through 13 of pregnancy caused a high incidence of abortion and/or embryonic resorp-tion in rats (Chatterjee and Harper, 1970); estrogen prevented the abor-tifacient effect of reserpine and chlorpromazine. Chlorpromazine and promazine have been shown to inhibit ovulation in rats and mice by interfering with the release of luteinizing hormone (Purshottam, Mason, and Pincus, 1961).

Information on the effect of tranquilizers on reproductive function in male animals is negligible. Gillette (1960) and Zimbardo and Barry (1958) found that the copulation rate, but not the percentage of rats that would copulate, was reduced by chlorpromazine administration. Addi-tions of promazine or chlorpromazine in concentrations below 200 $\mu g/ml$ of media suitable for prolonged storage were without effect on mobility of dog sperm, but higher levels of these drugs were spermicidal (Foote and Gray, 1963). Oral administration of chlorpromazine at a dose level of 4.4 mg/kg of body weight every other day or daily for 9 days did not reduce the libido of dogs (Foote and Gray, 1963). Mobility of the sperm collected was unaffected, and sperm output was inconsistently higher in the treated group.

BEHAVIORAL STRESS AND DISEASE IN ZOO ANIMALS

A classification of disease according to etiology includes the following categories: (1) genetic, (2) infectious, (3) parasitic, (4) diseases caused by physical agents, (5) nutritional, (6) autoimmune, (7) metabolic, (8) degenerative, and (9) neoplastic. Certain features of the stress reaction, especially increased adrenocortical function and involution of thymic and lymphoid tissue, theoretically increase an animal's susceptibility to infectious and parasitic disease. Some kinds of neoplasia, those initiated by viruses for example, may also be etiologically related to stress phe-nomena. Autoimmune, metabolic, and degenerative diseases in some in-stances, at least in a theoretical sense, are linked to intense and prolonged stress responses.

Exogenous cortisone, hydrocortisone, and ACTH decrease resistance to a variety of experimental infections in mammals. Such experiments are said to elucidate the mechanisms by which stress induces infectious dis-ease. The question arises whether or not the same events may occur as a result of increased endogenous secretion of the same or similar hormones.

A frequent criticism of experiments with exogenous hormones is that the results are pharmacological rather than physiological. In the same vein, Selye and his coworkers were able to induce hypertension, arthritis, arteriosclerosis, nephrosclerosis, and gastrointestinal ulcers experimentally by administering overdoses of hormones to animals sensitized by unilateral nephrectomy and a high dietary load of sodium chloride. Again, the criticisms leveled revolve around the question of whether spontaneous disease is very often the result of the natural stressors of everyday life.

The papers selected for citation in this review therefore meet two criteria: (1) diseases experimentally induced as a result of a natural behavioral stress, that is a stimulus that can reasonably be expected to be part of an animal's natural environment, and (2) diseases occurring in a captive wild animal.

Mortality from Injuries

Traumatic injuries are not classified as diseases but are often the result of behavioral stress and remain the most important single cause of death in zoo animals. Mortality figures for birds and mammals in the Philadelphia Zoo between 1951 and 1965 in Table 1 illustrate this point. Nearly

TABLE 1   Percentage of Total Deaths due to Injuries among Birds and Mammals in the Philadelphia Zoo, 1951–1965

	Mammals			Birds		
Year	No. Deaths	No. Attributed to Injury	%	No. Deaths	No. Attributed to Injury	%
1951	68	30	44	166	68	41
1952	44	18	41	128	57	45
1953	60	30	50	150	57	38
1954	72	29	40	120	71	59
1955	66	31	47	140	78	56
1956	65	41	63	153	75	49
1957	75	32	43	180	98	54
1958	46	15	33	197	105	53
1959	31	10	32	224	113	51
1960	42	17	40	165	69	42
1961	53	19	36	156	72	46
1962	48	14	29	158	72	46
1963	44	13	30	169	101	60
1964	34	6	18	160	54	34
1965	48	16	33	154	39	25
TOTAL	796	321	40	2,420	1,129	46

half of the deaths during this period were attributed to trauma. Some of these injuries were considered accidental or self-inflicted, but the majority were inflicted by cagemates. Conditions of captivity may not be responsible, but field studies show that intraspecific conflicts in nature seldom result in serious injuries (Wynne-Edwards, 1962).

Actually, most zoological gardens do not record the circumstances responsible for fatal injuries, so the number that can be attributed directly to behavioral stress is conjecture. Nevertheless, the losses are substantial and possibly could be prevented with attention to social factors. Several explanations for excessive mortality from intraspecific conflicts have been proposed but probably none are better than those offered by Charles Penrose in *Disease in Captive Wild Mammals and Birds* (Fox, 1923):

Captivity causes numerous physical and mental disarrangements. Unaccustomed, unnatural and unvaried food, change of climate and environment, physical and mental degeneration from disuse of muscle and brain, fear, ennui, nostalgia, lack of the exhilaration of chasing and being chased, unsatisfied sexual feeling—all react harmfully in the captive.

Penrose also commented on crowding, a factor much invoked as a primary fault in zoo management.

The size of the cage or pen has not as much effect upon the well-being of the animal as might be expected. Reptiles, birds, and mammals do as well in cages and pens of medium size as in very large ones . . . nor has a large enclosure been found perceptibly to diminish mortality from cagemates. The stronger will follow the weaker until he gets him, no matter what the enclosure.

In essence, the size of the cage is not as important as its complexity (e.g., presence of visual barriers) or the composition of the group. Sex ratio, age composition, behavior patterns are fully as important as population density.

Viral Infections

Laboratory mice exposed to psychological stress show an increased susceptibility to viral infection. Specifically, mice subjected to an "avoidance-learning" type of stressor are more susceptible to infection with the viruses of herpes simplex (Rasmussen, Marsh, and Brill, 1957), Coxsackie B (Johnsson, Lavender, and Marsh, 1959), and vesicular stomatitis (Jensen and Rasmussen, 1962). In general, increased susceptibility is attributed to depressed immunological function, a nonspecific response to stress. At present, there are no well-documented studies of the relationships between behavioral stress and viral infections in zoo animals; however, studies performed by Kalter (1973) are pertinent to the problem.

He reported the 5-month period following shipment of animals from Africa to be the time of highest virus isolation. This observation was attributed to the "stress" of traveling. It is not inconceivable that behavioral stress is equally capable of activating latent viruses. The list of recognized simian viruses now stands at 70+.

Shigellosis

Fatal ulcerative colitis associated with sudden alterations of the social environment has been described in four siamang gibbons (Stout and Snyder, 1969). The microscopic characteristics of the lesions in some respects were quite similar to those of idiopathic ulcerative colitis in man. Because ulcerative colitis in humans has been correlated with feelings of hopelessness and helplessness, often precipitated by real or fantasied losses or separations from loved ones (Engle, 1961), the occurrence of similar lesions in a nonhuman primate following emotional deprivation was reason for considerable interest.

During the summer of 1970 a study of the anthropoid apes in the Philadelphia Zoo was conducted by S. G. Souic, P. Claghorn, and myself to ascertain the status of *Shigella* bacteria. *S. flexneri,* types 1, 2 or both, were isolated from stools of eight of fifteen primates in a preliminary survey (Table 2). The distribution of *Shigella* organisms in the various

TABLE 2   Status of *Shigella flexneri* in Primate Groups at the Philadelphia Zoo in 1970

Group No.	Species	No. in Group	No. Positive
1	*Gorilla gorilla*	6	5
2	*Gorilla gorilla*	2	1
3	*Pongo pygmaeus*	2	1
4	*Pongo pygmaeus*	3	1
5	*Symphalangus syndactylus*	1	0
6	*Hylobates lar*	1	0
TOTAL		15	8

exhibition groups is illuminating and illustrates again the complexity of the epidemiological factors. Only 1 of the 15 animals in question showed signs of illness; a subadult female gorilla had exhibited intermittent bouts of diarrhea for several years. A male cagemate died with ulcerative colitis of undetermined etiology in 1967. The extremes in ages of the animals testing positive, a 3-year-old orangutan and a 53-year-old orangutan, again, are certainly a provocative discovery.

Spontaneous shigellosis in monkeys is characterized by a carrier state, with and without colonic mucosal abnormalities, as well as by the more familiar acute and often epidemic process (Ruch, 1959; Schneider *et al.*, 1960; Lapin and Yakovleva, 1963; Takasaka *et al.*, 1964; Ogawa *et al.*, 1964).

The existence of so many *Shigella* carriers in groups of apparently healthy anthropoids is reason enough to consider behavioral stress as a potentiating factor. The four gibbons with ulcerative colitis had one experience in common, emotional stress. Although not documented, it is well known that mortality is highest among newly imported primates. After adaptation to captivity the problems with shigellosis and other bacterial infections are not as common.

The question of virulence of various races of bacteria is a nagging problem. The possibility that the anthropoids in the Philadelphia Zoo are infected with avirulent strains of *S. flexneri* is a valid question and certainly should be explored. However, given the experimental studies of stress-induced viral, bacterial, and parasitic diseases, the most plausible explanation for sudden infection is a diminution of the immunological defense mechanisms of the host.

There are several other examples of ulcerative colitis in monkeys and apes among the postmortem records of the Penrose Research Laboratory. The similarity of this colonic disease to chronic idiopathic ulcerative colitis in man has not been established. It is generally agreed that a disease similar to the human form does not occur in animals. In view of the prevalence of *Shigella* organisms in zoo populations of primates, it is unlikely that such animals will prove useful models to elucidate mechanisms of pathogenesis unless the supposition that bacteria are not involved is in error.

Tuberculosis

Tobach and Bloch (1956) reported that the survival time of mice of both sexes suffering from acute tuberculosis decreased after infection under crowded conditions, whereas crowding after infection had essentially no effect on the course of chronic tuberculosis in females, but intensified the chronic disease in males.

Tuberculosis is a major cause of illness and death of primates in zoological gardens but unfortunately factors affecting susceptibility have been studied hardly at all. Fiennes (1970) states that the stresses of shipping, caging, and diet increase susceptibility of primates to tuberculosis, but no one else to my knowledge has mentioned the stress factor as being very important to the problem.

Amebiasis

Prior to 1935 *Entamoeba histolytica* was regarded as a pathogenic organism when present in the intestine of man. This belief has been gradually eroded by evidence that many individuals carry this protozoan without having intestinal disease. As a matter of interest, the epidemiology of this protozoan is quite analogous to that of the *Shigella* bacteria, factors determining the pathogenicity of either organism being largely unknown.

Experimental studies of amebic infections early in this century established the principle of lowered resistance as a prerequisite to the development of disease. Such factors as poor conditions, concomitant bacterial infection, other diseases, special intestinal flora, and poor nutrition lowered the resistance of the host and allowed the ameba to "invade" the tissue.

The records of the Penrose Research Laboratory list fourteen cases of amebiasis among mammals, the last case occurring in 1956. Eight monkeys, four black spider monkeys (*Ateles paniscus*), two Geoffroy's spider monkeys (*Ateles geoffroyi*), and two woolly monkeys (*Lagothrix lagotricha*), died of amebic dysentery in 1930. Deaths occurred between 1 and 122 days after the monkeys were received in the zoo, which suggests the introduction of a pathogenic amoeba from outside as the underlying epidemiological factor. A woolly monkey that died of amebiasis in 1938 was part of the 1930 importation. Altogether, six different species were involved—the Indian langur (*Presbytis entellus*), the Red Howler Monkey (*Aloutta seniculus*), and the long-haired spider monkey (*Ateles belzebuth*), in addition to those mentioned above. The exhibition periods ranged between 1 day and 18 months for 13 of the monkeys; the animal dying in 1938 was the notable exception since it had lived in the zoo for 8 years.

Postmortem records list 720 primates among the 6,570 mammals studied between 1901 and 1972; thus, the incidence of amebic disease recognized by postmortem examinations is 1.9 percent for primates alone and 0.2 percent for all mammals. Mortality from amebic disease, although a dreaded infection in man, then, should not be cause for much anxiety in the animal collection.

Again, I investigated the status of this organism among the anthropoid apes in the Philadelphia Zoo. Four groups of apes (see Table 3) were studied between 1968 and 1973. These were exhibition groups, with the animals living as cagemates. Pooled stool samples from each group were examined for *E. histolytica* once each year. Stool samples every year from each group were always positive (Table 3), again indicating the widespread distribution of this organism in the carrier state. Group one was comprised of subadult gorillas (transferred to another zoo in 1971),

TABLE 3   *Entamoeba histolytica* Recovered from Stools of Anthropoid Apes in the Philadelphia Zoo, 1968–1973

Group No.	Species	No. in Group	Consecutive Years Positive
1	*Gorilla gorilla*	2	1968–1971
2	*Gorilla gorilla*	6	1970–1973
3	*Pan troglodytes*	4	1969–1973
4	*Pongo pygmaeus*	3	1971–1973

group two of young gorillas, group three of subadult chimpanzees, and group four of young orangutans.

E. histolytica varies considerably in size, feeding habits, and pathogenicity, which is interpreted by some to mean that more than one species is included under this name. An alternate explanation is that size and feeding habits depend on the physicochemical environment of the gut and the intestinal flora, not genetic differences. Pathogenicity is thought to depend upon the organism's ability to invade the host's tissue. The confusion with regard to the problem of virulence or pathogenicity is further illustrated by the supposition that nonpathogenic, commensal species of *E. histolytica* suddenly acquire the property of invasiveness and become pathogenic.

In contrast *Entamoeba invadens* is considered a highly virulent ameba in reptiles (Geiman and Ratcliffe, 1936, Ratcliffe and Geiman, 1938). This species resembles *E. histolytica* morphologically and is responsible for periodically severe cases in the reptile collections of nearly all zoological gardens. Water snakes are apparently most susceptible to the disease, but lizards succumb occasionally. Turtles are apparently natural carriers of the organisms, hence a source of infection to snakes and reptiles in mixed exhibits, although recently snake-headed turtles (*Chelodina* spp.) and map turtles (*Graptemys* spp.) have died of this disease in the Philadelphia Zoo.

In retrospect, the principle of lowered resistance as a prerequisite of invasiveness for the amebae fits the concept of adaptive responses postulated by Selye. Concomitant bacterial infection and poor nutrition, for example, are potent stressors, thus lowered resistance to *E. histolytica* and *E. invadens* can be attributed to increased adrenocortical secretions and involution of lymphoid tissue. Specifically, the carbohydrate-active corticoids of the adrenal cortex suppress inflammation, phagocytosis, granulation, and antibody formation (Taubenhaus and Amromin, 1950; Selye, 1951; Dorfman, 1953; Dougherty, 1953; Taubenhaus, 1953; Dougherty and Schneebeli, 1955; and Kass, Kendrick, and Finland,

1955). The ability of the ameba to invade the intestinal tissue of the host is probably not the sole prerequisite for infectiveness. Defense mechanisms of the host may be more important. Suppression of inflammation, phagocytosis, and antibody formation may constitute the mechanism for the "lowered resistance" alluded to earlier.

Alteration of intestinal flora, suggested earlier as a predisposing factor for amebiasis, is also a consequence of alarming stimuli, ACTH, and cortisone. Pituitary and adrenocortical hormones affect the function of the mucosa of the stomach and intestinal tract (Baker and Abrams, 1954; Baker and Bridgman, 1954; and Gray and Ramsey, 1957).

Parasitic Disease

Either cortisone or ACTH increases the invasiveness of *Trichinella* larvae, by suppressing the defensive inflammatory response of the host's intestinal wall, and possibly by prolonging the sojourn of the adult females in the gut by suppressing immune responses to the worms (Stoner and Godwin, 1953). Grouping evidently stimulates a sufficient natural increase in the secretion of adrenal corticoids to produce similar effects, as this stimulus enhances the susceptibility of wild house mice to invasion by the larvae of *Trichinella spiralis* (Davis and Read, 1958). Davis and Read infected parenterally each mouse in their experiment with approximately 125 embryonated *Trichinella* larvae. Each mouse was maintained in a separate cage, but, from days 3 through 11 after infection, 11 of the mice were separated into two groups—one of five and the other of six—for 3 hours a day, while 11 others were left segregated. The mice were sacrificed the 15th day after infection and the gastrointestinal tracts digested and the larval worms recovered. Only three of the segregated mice were infected with an average of nine worms apiece, while all of the grouped mice were infected with an average of 32 worms apiece.

Amyloidosis

Amyloidosis is a disease with accumulations of abnormal glycoprotein around blood vessels of the spleen, kidneys, liver, pancreas, thyroids, parathyroids, adrenals, gonads, intestines and muscles. Etiology is uncertain, but its occurrence in horses repeatedly immunized for the production of diphtheria antitoxin, in human patients with long-standing infections such as tuberculosis, and in human patients with rheumatoid arthritis suggests that the disease may be an autoimmune disorder.

Cowan (1968a) reviewed the postmortem records of 1,698 birds from 22 orders and 76 families in the Philadelphia Zoo. The incidence of amyloidosis among susceptible families ranged from a high of 45 percent (262 of 578) among Anatidae to a low of 0.6 percent among Psittacidae.

The relationship of amyloidosis to chronic disease was problematic. Tuberculous lesions found were often small and at an early stage of development, especially in birds with far advanced amyloidosis. Aspergillosis, a mould disease, was seen as a chronic granulomatous disease, especially in penguins, but it was more often only a secondary infection in extremely debilitated birds. Attempts to separate the disease entities into primary or secondary forms were considered specious.

Cowan (1968b) also analyzed the postmortem records of the Anatidae with respect to the influence of genetic and specific predilection, age, sex incidence, population density, behavior patterns, and adaptability to the zoo environment. Within 13 genera proportionate morbidity ranged from 29 percent for the genera *Tadorna* and *Dendrocygna* to 92 percent for the genus *Chloephoga*. Other genera were also affected, but numbers were too small to analyze statistically. Cowan surmised that amyloidosis appeared most commonly in birds subjected for prolonged periods to environmental stressors.

Cowan describes the whistling ducks, *Dendrocygna,* as shy, nonaggressive, and easily domesticated. Their incidence of amyloidosis is correspondingly low. The genera *Chloephoga* and *Tadorna* belong to the same tribe, Tadornini, which is composed mainly of aggressive, competitive birds. However, while *Chloephoga* is a South American genus exclusively, *Tadorna* is widely distributed. In addition, the less aggressive, more gregarious, common sheldrake, *T. tadorna,* makes up more than half of the *Tadorna* group on exhibit in Philadelphia. With this interpretation of behavior patterns, Cowan suggested that susceptibility to amyloid disease was related less to genetic predisposition and more to inherent properties of adaptability.

A large proportion of the waterfowl in the Philadelphia Zoo exists as pinioned flocks on a large lake (approximately 2.5 acres) and a series of connecting ponds called bird valley. The waterfowl collection was expanded from roughly 200 birds on exhibit in 1945 to 800 in 1965. The average survival time dropped during the same period from almost 120 months between 1941 and 1945 to less than 80 months between 1951 and 1965. Thus, the incidence of amyloidosis is directly (positively) correlated with population density in the Philadelphia Zoo and bears no relationship to age.

Cowan's analysis of the Philadelphia data might have been dismissed as interesting but not definitive, had it not been for his follow-up laboratory studies (Cowan and Johnson, 1970). He studied white Pekin ducks, isolated, paired, and in flocks of eight. Amyloidosis developed in 21 percent of isolated ducks, in 42 percent of paired ducks, and in 71 percent of grouped ducks. These birds were otherwise healthy and free of parasites and only 265 days old when the study was terminated. The correlation of amyloid disease with population density was unequivocal.

Cowan's results are consistent with studies utilizing inbred strains of laboratory mice. Injections of ACTH or cortisone produce very high incidences of amyloidosis (Latvalahti, 1953). Also, chronic stress in the form of electric shocks produces the disease in mice (Hall, Cross, and Hale, 1960).

The mechanisms responsible for the production of the glycoprotein substance are unknown, but the scheme proposed by Cohen (1967) illustrates how adaptive responses to stress might be involved. In this scheme, a wide variety of stimuli, act upon a "stem cell," which gives rise to reticuloendothelial cells and plasma cells. The reticuloendothelial cells produce amyloid fibrils *in situ*, which in turn are deposited in normal or altered ground substance.

Arteriosclerosis

Emotional stress has long been suspected of causing or aggravating arteriosclerosis in man (Wolf, 1964). Studies of the relationship between behavioral stress and arteriosclerosis among captive animals have been encouraged because several interacting factors, such as dietary lipids, can be either controlled or monitored. In this respect, the studies of arteriosclerosis in the Philadelphia Zoo have been especially germane, as formulated feeding started there in 1935.

The basic omnivore diet contains less than 9-percent fat. Moreover, three items—cottonseed oil, cod liver oil, and chicken parts—which make up about three-fourths of the fat complex, are high in polyunsaturated fatty acids (68–85 percent). When the meat additive is chicken parts, dietary cholesterol is almost nil. In short, the omnivore diet at Philadelphia, supplemented with oranges, carrots, kale, and cabbage, is close to the kind of diet advocated for humans to prevent "hardening of the arteries." Arteriosclerosis still occurs at the Philadelphia Zoo, and mammals and birds continue to have "heart attacks." However, to say that low-fat diets and polyunsaturated fatty acids were, therefore, ineffective in preventing this disease would be unfair.

Ratcliffe and Cronin (1958) reported arteriosclerosis in less than 3 percent of autopsies on mammals and birds at the Philadelphia Zoo from 1901 through 1932, with the frequency of occurrence increasing to 20 percent by 1955. They concluded that part of the increase was related to age; the new diets contributing chiefly by improving chances for survival. Because the increased frequency of occurrence was most pronounced from 1945 through 1955 and corresponded to increases in population density, these investigators suggested that social pressure, through an imbalance in adrenal secretion, had become a major factor in the increasing frequency of this disease.

Dietary improvements in the Philadelphia Zoo were followed by significant changes in the characteristics and locations of arterial lesions (Ratcliffe, Yerasimides, and Elliott, 1960). The large, soft atheromata of the proximal aorta and brachiocephalic arteries characteristic of birds before 1935 were replaced by smaller, more compact lesions, now usually of the abdominal aorta. Lesions of the aorta and its large branches were apparently never prevalent among mammals, even before 1923 (Fox, 1923); but now, in 1973, significant lesions of the aorta and large proximal coronary arteries of the heart are practically nonexistent in this class of animal. Complications of the atheromatous lesions, such as ulcerations and dissecting aneurisms, so common in the human form of the disease, were rarely observed in the Philadelphia animals.

The most significant lesions now are apparently widespread in small and intermediate sized vessels at all levels of the arterial system in both mammals and birds. Arteriosclerosis in these locations develop through reorientation, proliferation, and possibly migration of pleomorphic smooth muscle cells in the arterial wall, combined with increased formation of mucopolysaccharides, and elastic and collagen fibers. However, even though larger arteries are not severely involved and atheromata are small and uncomplicated, spontaneous myocardial infarction and fibrosis, and cerebral accidents, are still important causes of death among the birds and mammals of the Philadelphia Zoo.

The omnivore diet supplemented with carrots, kale, and cabbage has been tested on captive woodchucks (*Marmota monata*) over a 13-year period. Preliminary results of this study were reported after 8 years (Snyder and Ratcliffe, 1969). Now, postmortem studies total 65, but atheromata of the aorta and of the proximal coronary arteries of the heart have not occurred, even though a substantial proportion of the experimental subjects were between 7 and 9 years old at death. In spite of the absence of macroscopic atheromata, these investigations have documented four cases of myocardial infarction (Figure 1) and five cerebral accidents due to rupture of arteriosclerotic cerebral arteries (Figure 2). Obviously, lesions of the smaller vessels can produce infarctions and strokes. In general, low-fat diets and polyunsaturated fatty acids are effective in reducing or preventing atheromata of the large arteries in captive birds and mammals, but whether dietary factors are involved in the pathogenesis of small vessel disease has not been determined.

Controlled experiments with chickens again suggest that social factors influence the course of arteriosclerosis of the intramural arteries and also demonstrated that arteriosclerosis of the intramural vessels of avian hearts

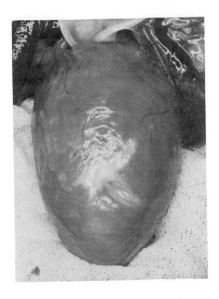

Figure 1
Anterior surface of the heart of a male woodchuck dead at 74 months of age. The white fibrous scar is the result of a previous transmural infarct of the wall of the left ventricle.

causes necrosis of the myocardium (Ratcliffe and Snyder, 1964, 1965, 1967). The male chickens always had more severe arteriosclerosis regardless of age or population density. As a matter of interest, males caged alone had significantly more severe arterial disease than those in large

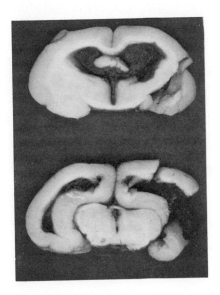

Figure 2
Cut sections of the brain of a male woodchuck dead at 36 months. Clotted blood in the lateral ventricles originated from a ruptured cerebral artery.

groups, which was also true of the experiments conducted by Katz and Pick (1961). Their birds were fed cholesterol. Arterial atheromatosis accelerated when the experimental subjects, accustomed to crowded social conditions, were kept in isolation.

The physiological mechanisms that might account for the correlations between behavioral stress and increased severity of arteriosclerosis have not been elucidated; however, the process of lipid mobilization, as well as the concentration of fats and cholesterol in the blood serum, are implicated because of the presumed relationships of altered fat metabolism to the development of arteriosclerosis. The catecholamine hormones promote lipolysis as do impulses from the sympathetic nerves. Thus, psychologically meaningful experiences in man are accompanied by sharp, often sustained, changes in the process of lipid mobilization as well as in the concentration of cholesterol and lipoproteins in the blood. Exogenous catecholamines, furthermore, have direct injurious effects on arterial smooth muscle and heart muscle (Reichenbach and Benditt, 1970; Selye, 1970).

SUMMARY AND CONCLUSIONS

Endocrine and physiological adaptive mechanisms are unquestionably important in enabling birds and mammals to meet and counteract their constantly changing environment and the vicissitudes of daily existence. These complex regulatory systems together with the autonomic and parasympathetic nervous systems control the internal environment of vertebrate animals within narrow limits. Stress exists when regulatory systems require more energy than is available to the animal, which is the case when environmental contingencies are severe and prolonged. Internal organismal stability is achieved by sacrificing functions less immediately vital to the individual, such as growth, reproducton, and resistance to infectious disease and parasitism.

William Cannon long ago described acute stress reactions of mammals to real or symbolic physical danger, including increased secretion of adrenaline, shortening of the clotting time of the blood, a rise in metabolic rate, tachycardia, increased cardiac output, and hypertension. Cannon suggested that chronic functional disturbances could develop when these reactions occur in inappropriate situations or cannot be extinguished by the release of physical activity. The endocrine or hormonal adjustments that occur during stress are emphasized particularly as being useful to the animal in maintaining homeostasis. Hans Selye proposed the hypothesis that the adaptive mechanisms called into operation may derail and cause disease or interfere with reproductive function if the noxious stimuli are prolonged and unrelieved.

Caged wild birds and mammals, faced with a number of problems not encountered in their natural habitats, are invariably more excitable than either their domestic or laboratory-reared cousins and infinitely more difficult to handle and restrain. A number of factors related to behavior are inherent features of the zoo and potentially capable of inducing stress reactions. Most behavioral interactions are termed social, i.e., involving interactions between animals, but responses to handling and restraint by humans are equally important in zoo management.

Drugs used to anesthesize, immobilize, or tranquilize animals can have direct pharmacological effects on hormonal and physiological systems that must be considered by investigators measuring stress responses or collecting so-called baseline data. Anesthetics render an animal unconscious, thus eliminating sensory input and stress reactions, but immobilizing chemicals usually only paralyze without interfering with visual and auditory acuity and consciousness. Inability to move generates feelings of fear and frustrations that are potent emotional stressors and difficult to counteract.

Certain features of the stress reaction, especially increased adrenocortical function and involution of thymic and lymphoid tissue, increase an animal's susceptibility to infectious and parasitic disease. Cases of fatal ulcerative colitis in siamang gibbons associated with sudden alterations of the social environment can be explained as a consequence of such responses. Stress of sufficient intensity to diminish immunological defenses is a common sequelae of animal transporting procedures, thus mortality from infectious and parasitic diseases is always more of a problem among new arrivals in the zoo.

Data collected in the Philadelphia Zoo show a high prevalence of *Shigella flexneri* and *Entamoeba histolytica* in the stools of the resident anthropoid apes. The available evidence suggests that well-nourished mammals acclimated to the zoo environment are resistant to these pathogenic agents. Conditions leading to stress are believed to lower the host's defenses sufficiently to allow these intestinal pathogens to invade the tissue. Pathogenicity or virulence of the bacteria and protozoa, then, may not be the key to infectiveness.

Injected ACTH and cortisone induce amyloidosis in laboratory mice; thus, the correlation of amyloid disease with population density in waterfowl of the Philadelphia Zoo has some experimental basis. Furthermore, amyloidosis has appeared most commonly in birds subjected for prolonged periods to environmental stressors. Follow-up laboratory studies with white Pekin ducks provided more data. Amyloidosis developed in 21 percent of isolated ducks and in 71 percent of grouped ducks within 265 days. Crowding was the only experimental variable.

Emotional stress has long been suspected of causing or aggravating arteriosclerosis in man. Studies of arteriosclerosis and behavioral stress among confined birds and mammals provide additional evidence for this idea. Again, features of chronic stress reactions, especially those elements affecting the process of lipid mobilization and concentrations of fats and cholesterol in the blood serum, are potentially capable of producing injury to or disrupting the integrity of the arterial wall. An important question to be answered is whether cholesterol and other lipid compounds initiate arterial lesions or simply appear after the stage is set by other factors. In either case, the concept of stress is a helpful guide for research.

The data and ideas presented in this report represent only a portion of the pertinent material accumulated from zoo research; nevertheless certain conclusions are evident. Judgments about the nature of biological phenomena obtained from one species may not apply to others. In reality it is the subtle variations of biological processes in different organisms that give clues to the solution of complex problems of nature. This technique is called comparative study. Research requiring the comparative approach or long-term control of diet is particularly sutied to zoological gardens. Zoos can be utilized for biomedical and biological research without interfering with the primary functions of these institutions.

REFERENCES

Ader, R. 1965. Effects of early experience and differential housing on behavior and susceptibility to gastric erosions in the rat. J. Comp. Physiol. Psychol. 60:233–238.

Ader, R. 1966. Frequency of stimulation during early life and subsequent emotionality in the rat. Psychol. Rep. 18:695–701.

Ader, R. 1969. Adrenocortical function and the measurement of "emotionality." Ann. N.Y. Acad. Sci. 159:791–805.

Ader, R. and S. B. Friedman. 1965. Psychological factors and susceptibility to disease in animals. In Symposium on Medical Aspects of Stress in the Military Climate. Walter Reed Army Institute of Research, Washington, D.C.

Allen, G. M. 1939. A checklist of African mammals. Bull. Mus. Comp. Zool., Vol. 83.

Allen, T. J. 1970. Immobilization of white-tailed deer with succinylcholine chloride and hyaluronidase. J. Wildl. Manage. 34:207–209.

Bailey, J. L., Jr. 1939. Physiological group differentiation in Lymnaea columella. Am. J. Hyg. Monogr. Ser. 14:1–133.

Baker, B. L., and G. D. Abrams. 1954. Effect of hypophysectomy on the cytology of the fundic glands of the stomach and on the secretion of pepsin. Am. J. Physiol. 177:409–412.

Baker, B. L., and R. M. Bridgman. 1954. The histology of the gastrointestinal mucosa (rat) after adrenalectomy and administration of adrenocortical hormones. Am. J. Anat. 94:363–397.

Barrett, A. M., and M. A. Stockham. 1965. The response of the pituitary–adrenal system to a stressful stimulus: The effect of conditioning and pentobarbitone treatment. J. Endocrinol. 33:145–152.

Bernstein, L., and H. Elrick. 1957. The handling of experimental animals as a control factor in animal research—A review. Metabolism 6:479–482.

Cannon, W. B. 1929. Bodily changes in pain, hunger, fear, and rage. C. T. Branford Co., Boston.

Cannon, W. B., and D. Rapport. 1921. Studies on the conditions of activity in endocrine glands. VII. The reflex center for adrenal secretion and its response to excitatory and inhibitory influences. Am. J. Physiol. 58:338–352.

Chatterjee, A. 1965. Ovarian response to administration of "PMS" in chlorpromazined and reserpinized female rats. Naturwissenschaften 52:309.

Chatterjee, A. 1966. The role of cortisone in the prevention of gonadal inhibition in chlorpromazined rats. Acta Anat. 65:606–609.

Chatterjee, A., and M. J. K. Harper. 1970. Interruption of implantation and gestation in rats by reserpine, chlorpromazine and ACTH: Possible mode of action. Endocrine 87:966–969.

Christian, J. J. 1955a. Effect of population size on the weights of the reproductive organs of white mice. Am. J. Physiol. 181:477–480.

Christian, J. J. 1955b. Effect of population size on the adrenal glands and reproductive organs of male mice in populations of fixed size. Am. J. Physiol. 182:292–300.

Christian, J. J. 1956. Adrenal and reproductive responses to population size in mice from freely growing populations. Ecology 37:258–273.

Christian, J. J. 1957. Reserpine suppression of density-dependent adrenal hypertrophy and reproductive hypoendocrinism in populations of male mice. Am. J. Physiol. 187:353–356.

Christian, J. J. 1959a. The roles of endocrine and behavioral factors in the growth of mammalian populations. Pages 71–97 in Proceedings, Columbia University Symposium on Comparative Endocrinology. John Wiley & Sons, New York.

Christian, J. J. 1959b. Control of population growth in rodents by interplay between population density and endocrine physiology. Wildl. Dis. 1:1–38.

Christian, J. J. 1959c. Adrenocortical, splenic, and reproductive responses to inanition and to grouping. Endocrine 65:189–196.

Christian, J. J., and D. E. Davis. 1955. Reduction of adrenal weight in rodents by reducing population size. Trans. N. Am. Wildl. Conf. 20:177–189.

Christian, J. J., and D. E. Davis. 1956. The relationship between adrenal weight and population status in Norway rats. J. Mammal. 37:475–486.

Christian, J. J., and C. D. LeMunyan. 1958. Adverse effects of crowding on lactation and reproduction of mice and two generations of their offspring. Endocrine 63:517–529.

Christian, J. J., and H. L. Ratcliffe. 1952. Shock disease in captive wild mammals. Am. J. Pathol. 28:725–737.

Christian, J. J., and H. O. Williamson. 1958. Effect of crowding on experimental granuloma formation in mice. Proc. Soc. Exp. Biol. Med. 99:385–387.

Cohen, A. S. 1967. Amyloidosis. N. Engl. J. Med. 277:522–529, 574–583, 628–638.

Coolidge, H. L. 1929. A revision of the genus *Gorilla*. Mem. Mus. Comp. Zool. 50:295–381.

Cowan, D. F. 1968a. Avian amyloidosis. I. General incidence in zoo birds. Pathol. Vet. 5:51–58.

Cowan, D. F. 1968b. Avian amyloidosis. II. Incidence and contributing factors in the family Anatidae. Pathol. Vet. 5:59–66.

Cowan, D. F., and W. C. Johnson. 1970. Amyloidosis in the white Pekin duck. I. Relation to social environmental stress. Lab. Invest. 23:551–555.

Crandall, L. S. 1964. The management of wild mammals in captivity. University of Chicago Press, Chicago and London.

Crockford, J. A., F. A. Hayes, J. H. Jenkins, and S. D. Feurt. 1957. Nicotine salicylate for capturing deer. J. Wildl. Manage. 21:213–220.

Davis, D. E. 1951a. The relation between level of population and pregnancy of Norway rats. Ecology 32:459–461.

Davis, D. E. 1951b. The relation between level of population and size and sex of Norway rats. Ecology 32:462–464.

Davis, D. E., and C. P. Read. 1958. Effect of behavior on development of resistance to trichinosis. Proc. Soc. Exp. Biol. Med. 99:269–272.

Dobzhansky, T., and C. Epling. 1944. Contributions to the genetics, taxonomy, and ecology of Drosophila pseudoobscura and its relatives. Publ. Carnegie Inst. Wash. 554:1–183.

Domino, E. F. 1964. Neurobiology of phencyclidine (Sernyl), a drug with an unusual spectrum of pharmacological activity. Int. Rev. Neurobiol. 6:303–347.

Dorfman, A. 1953. The effects of adrenal hormones on connective tissue. Ann. N.Y. Acad. Sci. 56:698–703.

Dougherty, T. F. 1953. Some observations on mechanisms of corticosteroid action on inflammation and immunologic processes. Ann. N.Y. Acad. Sci. 56:748–756.

Dougherty, T. F., and G. L. Schneebeli. 1955. The use of steroids as anti-inflammatory agents. Ann. N.Y. Acad. Sci. 61:328–348.

DuShane, G. P., and C. Hutchinson. 1944. Differences in size and development rate between eastern and midwestern embryos of Ambystoma maculatum. Ecology 25:414–423.

Eibl-Eibesfeldt, I. 1970. Ethology, the biology of behavior. Holt, Rinehart and Winston, New York.

Engle, G. L. 1961. Biologic and psychologic features of the ulcerative colitis patient. Gastroenterology 40:313–322.

Fiennes, R. N. T-W. 1970. Pages 149–154 in H. Balner and W. I. B. Beveridge, eds. Infections and immunosuppression in subhuman primates. Munksgaard, Copenhagen.

Flickinger, G. L., Jr. 1961. Effect of grouping on adrenals and gonads of chickens. Gen. Comp. Endocrinol. 1:332–340.

Flickinger, G. L., Jr. 1966. Response of the testes to social interaction among grouped chickens. Gen. Comp. Endocrinol. 6:89–98.

Foote, R. H., and L. C. Gray. 1963. Effect of tranquilizers on libido, sperm production and in vitro sperm survival in dogs. Proc. Soc. Exp. Biol. Med. 114:396–398.

Fox, H. 1923. Disease in captive wild mammals and birds. Lippincott, Philadelphia.

Fox, K. A., and R. L. Snyder. 1969. Effect of sustained low doses of diazepam on aggression and mortality in grouped male mice. J. Comp. Physiol. Psychol. 69:663–666.

Friedman, S. B., and R. Ader. 1967. Adrenal cortical response to novelty and noxious stimulation. Neuroendocrinol. 2:209–212.

Fulton, B. B. 1933. Inheritance of song in hybrids of two subspecies of Nemobius fasciatus (Orthoptera). Ann. Entomol. Soc. Am., 26:368–376.

Gaunt, R., A. A. Renzi, N. Antonchak, G. J. Miller, and M. Gillman. 1954.

Endocrine aspects of the pharmacology of reserpine. Ann. N.Y. Acad. Sci. 59: 22–35.

Geiman, Q. M., and H. L. Ratcliffe. 1936. Morphology and life-cycle of an amoeba producing amoebiasis in reptiles. Parasitology 28:208–228.

Gillette, E. 1960. Effects of chlorpromazine and D-lysergic acid diethylamide on sex behavior of male rats. Proc. Soc. Exp. Biol. Med. 103: 392–394.

Giuliani, G., M. Motta, and L. Martini. 1966. Reserpine and corticotrophin secretion. Acta Endocrinol. 51:203–209.

Gold, E. M., and W. F. Ganong. 1967. Effects of drugs on neuroendocrine processes. Pages 377–437 in L. Martini and W. F. Ganong, eds. Neuroendocrinology, Vol. 2. Academic Press, New York and London.

Goldman, J. R. 1965. Conditioned emotionality in the rat as a function of stress in infancy. Anim. Behav. 13:434–442.

Gray, S. J., and C. G. Ramsey. 1957. Adrenal influence upon the stomach and the gastric responses to stress. Recent Progr. Hormone Res. 13:583–617.

Hall, C. E., E. Cross, and O. Hale. 1960. Amyloidosis and other pathologic changes in mice exposed to chronic stress. Tex. Rep. Biol. Med. 18:205–213.

Hall, T. C., E. B. Taft, W. H. Baker, and J. C. Aub. 1953. A preliminary report on the use of Flexedil to produce paralysis in the white-tailed deer. J. Wildl. Manage. 17:516–520.

Harthoorn, A. M. 1962. On the use of phencyclidine for narcosis in the larger animals. Vet. Rec. 74:410–411.

Harthoorn, A. M. 1965. Application of pharmacological and physiological principles in restraint of wild animals. Wildl. Monogr. 14:8–78.

Harthoorn, A. M., and J. Bligh. 1965. The use of a new oripavine derivative with potent morphine-like activity for the restraint of hoofed wild animals. Res. Vet. Sci. 6:290–299.

Hatch, A., T. Balazs, G. S. Wiberg, and H. C. Grice. 1963. Long-term isolation stress in rats. Science 142:507.

Hediger, H. 1950. Wild animals in captivity: An outline of the biology of zoological gardens. Butterworths, London. [New Ed. 1964, Dover Publications, Inc., New York.]

Hediger, H. 1969. Man and animals in the zoo. Seymour Lawrence/Delacorte Press, New York.

Hinde, R. A. 1966. Animal behavior: A synthesis of ethology and comparative psychology. McGraw-Hill, New York.

Hofmeyr, C. F. B. 1960. Some observations on the use of succinylcholine chloride (suxamethonium) in horses with particular reference to the effect on the heart. J. S. Afr. Vet. Med. Assoc. 31:251–259.

Jensen, M. M., and A. F. Rasmussen. 1962. Biphasic changes in leucocyte counts and susceptibility to vesicular stomatitis virus in sound-stressed mice. Bacteriol. Proc. 149.

Johnsson, T., J. F. Lavender, and J. T. Marsh. 1959. The influence of avoidance-learning stress on resistance to Coxsackie virus in mice. Fed. Proc. 18:575.

Kalter, S. S. 1973. Virus research. Pages 61–165 in G. H. Bourne, ed. Nonhuman primates and medical research. Academic Press, New York and London.

Kass, E. H., M. I. Kendrick, and M. Finland. 1955. Effects of corticosterone, hydrocortisone and corticotrophin on production of antibodies in rabbits. J. Exp. Med. 102:767–774.

Katz, L. N., and R. Pick. 1961. Experimental atherosclerosis as observed in the chicken. J. Atheroscler. Res. 1:93–100.

Kroll, W. R. 1963. Experience with Sernylan in zoo animals. Int. Zoo Yearbk. 4:131–141.

Krushak, D. H., and W. V. Hartwell. 1968. Effect of phencyclidine on serum transaminase in chimpanzees. J. Am. Vet. Med. Assoc. 153:866–867.

Kuo, Z. Y. 1932. Antogeny of embryonic behavior in Aves. J. Exp. Biol. 61:395–430, 453–489.

Lapin, B. A., and L. A. Yakovleva. 1963. Comparative pathology in monkeys. Charles C Thomas, Springfield, Ill.

Latvalahti, J. 1953. Experimental studies on the influence of certain hormones on the development of amyloidosis. Acta Endocrinol. 14 (Suppl. 16):3.

Lehrman, D. S. 1953. A critique of Konrad Lorenz's theory of instinctive behavior. Q. Rev. Biol. 28:337–363.

Levy, L., D. E. Cameron, and R. C. B. Aitken. 1960. Observation on two psycho-tomimetic drugs of piperidine derivation—CI 395 (Sernyl) and CI 400. Am. J. Psychiatr. 116:843–844.

Lorenz, K. 1961. Phylogenetische Anpassung und adaptive Modifikation des Ver-haltens. Z. Tierpsychol. 18:139–187.

Lorenz, K. 1970. Studies in animal and human behavior. Harvard University Press, Cambridge, Mass.

Meites, J., C. S. Nicoll, and P. K. Talwalker. 1963. *In* A. V. Nalbandov, ed. Advances in neuroendocrinology, Vol. 2. Academic Press, New York and London.

Melby, E. C., and H. J. Baker. 1965. Phencyclidine for analgesia and anesthesia in simian primates. J. Am. Vet. Assoc. 147:1068–1072.

Ogawa, H., R. Takahashi, S. Honjo, M. Takasaka, T. Fujiwara, K. Audoo, M. Nakagawa, T. Muto, and K. Imaizumi. 1964. Shigellosis in cynomolgus monkeys (*Macaca irus*). III. Histopathological studies on natural and experimental shigellosis. Jap. J. Med. Sci. Biol. 17:321–332.

Olds, M. E., and G. Baldrighi. 1968. Effects of meprobamate, chlordiazepoxide, diazepam, and sodium pentobarbital on visually evoked responses in the tectoteg-mental area of the rat. Int. J. Neuropharm. 7:231–239.

Psychoyos, A. 1963. A study of the hormonal requirements for ovum implantation in the rat, by means of delayed nidation-inducing substances (chlorpromazine, trifluorperazine). J. Endocrinol. 27:337–343.

Psychoyos, A. 1966. *In* G. E. W. Wolstenholme and M. O'Connor, eds. Egg implantation. Ciba Foundation Study Group, Little, Brown & Co.

Purshottam, N., M. M. Mason, and G. Pincus. 1961. Induced ovulation in the mouse and the measurement of its inhibition. Fert. Steril. 12:346–352.

Randall, L. O., G. A. Heise, W. Schallek, R. E. Bogdon, R. Banziger, A. Boris, R. A. Moe, and W. B. Abrams. 1961. Pharmacological and clinical studies on Valium (TM): A new psychotherapeutic agent of the benzodiazepine class. Cur. Therapeutic Res. 3:405–425.

Rasmussen, A. F., Jr., J. T. Marsh, and N. G. Brill. 1957. Increased susceptibility to herpes simplex in mice subjected to avoidance-learning stress or restraint. Proc. Soc. Exp. Biol. Med. 96:183–189.

Ratcliffe, H. L. 1968. Environment, behavior and disease. Pages 161–228 *in* E. Stellar and J. M. Sprague, eds. Progress in physiological psychology, Vol. 2. Academic Press, New York and London.

Ratcliffe, H. L., and M. T. I. Cronin. 1958. Changing frequency of arteriosclerosis in mammals and birds at the Philadelphia Zoological Garden. Review of autopsy records. Circulation 18:41–52.

Ratcliffe, H. L., and Q. M. Geiman. 1938. Spontaneous and experimental amebic infection in reptiles. Arch. Pathol. 25:160–184.

Ratcliffe, H. L., and R. L. Snyder. 1964. Myocardial infarction: A response to social interaction among chickens. Science 144:425–426.

Ratcliffe, H. L., and R. L. Snyder. 1965. Coronary arterial lesions in chickens: Origin and rates of development in relation to sex and social factors. Circ. Res. 17:403–413.

Ratcliffe, H. L., and R. L. Snyder. 1967. Arteriosclerotic stenosis of the intramural coronary arteries of chickens: Further evidence of a relation to social factors. Br. J. Exp. Pathol. 48:357–365.

Ratcliffe, H. L., T. G. Yerasimides, and G. A. Elliott. 1960. Changes in the character and location of arterial lesions in mammals and birds in the Philadelphia Zoological Garden. Circulation 21:730–738.

Reichenbach, P. D., and E. P. Benditt. 1970. Catecholamines and cardiomyopathy: The pathogenesis and potential importance of myofibrillar degeneration. Human Pathol. 1:125–150.

Ricker, W. E. 1938. "Residual" and Kokanec salmon in Cultus Lake. J. Fish. Res. Board Can. 4:192–218.

Ropartz, P. 1968. The relation between olfactory stimulation and aggressive behavior in mice. Anim. Behav. 16:97–100.

Ruch, T. C. 1959. Diseases of laboratory primates. W. B. Saunders Company, Philadelphia.

Schallek, W., and A. Kuehn. 1965. Effects of benzodiazepines on spontaneous EEG and arousal responses of cat. Progr. Brain Res. 18:231–238.

Schleiter, H., and K. H. Casper. 1961. Pathological changes in the heart of horses cast with and without suxamethonium. Vet. Med. 16:835–837.

Schneider, N. J., E. C. Prother, A. L. Lewis, G. E. Scatterday, and A. V. Hardy. 1960. Enteric bacteriological studies in a large colony of primates. Ann. N.Y. Acad. Sci. 85:935–941.

Schultz, A. H. 1934. Some distinguishing characteristics of the mountain gorilla. J. Mammal. 15:51–61.

Seal, U. S., A. W. Erickson, and J. G. Mayo. 1970. Drug immobilisation of the carnivora. Int. Zoo Yearbk. 10:157–170.

Selye, H. 1951. The influence of STH, ACTH and cortisone upon resistance to infection. Can. Med. Assoc. J. 64:489–494.

Selye, H. 1959. Perspectives in stress research. Perspect. Biol. Med. 2:403–416.

Selye, H. 1970. Experimental cardiovascular diseases. Springer-Verlag, New York, Heidelberg, Berlin.

Siegel, H. S. 1959. Egg production characteristics and adrenal function in White Leghorns confined at different floor space levels. Poult. Sci. 38:893–898.

Snyder, R. L. 1961. Evolution and integration of mechanisms that regulate population growth. Proc. Natl. Acad. Sci. (USA) 47:449–455.

Snyder, R. L. 1967. Fertility and reproductive performance of grouped male mice. Pages 458–472 in K. Benirschke, ed. Symposium on comparative aspects of reproductive failure. Springer-Verlag, New York.

Snyder, R. L. 1968. Reproduction and population pressures. Pages 119–160 in E.

Stellar and J. M. Sprague, eds. Progress in physiological psychology, Vol. 2. Academic Press, New York and London.

Snyder, R. L., and H. L. Ratcliffe. 1969. *Marmota monax*: A model for studies of cardiovascular, cerebrovascular and neoplastic disease. Acta Zool. Pathol. Antverpiensia 48:265–273.

Stevenson, D. E., and L. W. Hall. 1959. Pharmacological effects of suxamethonium. Vet. Rec. 71:818.

Stoner, R. D., and J. T. Godwin. 1953. The effects of ACTH and cortisone upon susceptibility to trichinosis in mice. Am. J. Pathol. 29:943–950.

Stout, C., and R. L. Snyder. 1969. Ulcerative colitis-like lesions in siamang gibbons. Gastroenterology 57:256–261.

Stowe, C. M. 1955. The curariform effect of succinyl choline in the equine and bovine species—Preliminary report. Cornell Vet. 45:193.

Takasaka, M., S. Honjo, T. Jujiwara, T. Hagiwara, H. Ogawa, and K. Imaizumi. 1964. Shigellosis in cynomolgus monkeys (*Macaca irus*). I. Epidemiological surveys on *Shigella* infection rate. Jap. J. Med. Sci. Biol. 17:259–265.

Talwalker, P. K., J. Meites, C. S. Nicoll, and T. F. Hopkins. 1960. Effects of chlorpromazine on mammary glands of rat. Am. J. Physiol. 199:1073–1076.

Taubenhaus, M. 1953. The influence of cortisone upon granulation tissue and its synergism and antagonism to other hormones. Ann. N.Y. Acad. Sci. 55:666–673.

Taubenhaus, M., and D. Amromin. 1950. The effects of the hypophysis, thyroid, sex steroids, and the adrenal cortex upon granulation tissue. J. Lab. Clin. Med. 36: 7–18.

Tavernor, W. D. 1960. The effect of succinyl choline chloride on the heart of the horse; clinical and pathological aspects. Vet. Rec. 72:269–572.

Tobach, E., and H. Bloch. 1956. Effect of stress by crowding prior to and following tuberculous infection. Am. J. Physiol. 187:399–402.

Valzelli, L., E. Giacalone, and S. Garattini. 1967. Pharmacological control of aggressive behavior in mice. Eur. J. Pharm. 2: 144–146.

Vessey, S. 1967. Effects of chlorpromazine on aggression in laboratory populations of wild house mice. Ecology 48:367–376.

Vondruska, J. F. 1965. Phencyclidine anesthesia in baboons. J. Am. Vet. Med. Assoc. 147:1073–1076.

Watson, J. B. 1919. Psychology from the standpoint of a behaviorist. Lippincott, Philadelphia.

Weininger, O. 1954. Physiological damage under emotional stress as a function of early experience. Science 119:285–286.

Willoughby, D. P. 1950. The gorilla: Largest living primate. Sci. Mon. 70:48–57.

Wolf, S. 1964. Psychological and social factors in cardiovascular disease. *In* The heart and circulation, Vol. I, part 1. Second National Conference on Cardiovascular Diseases Research, Washington, D.C.

Wynne-Edwards, V. C. 1962. Animal dispersion in relation to social behavior. Hafner, New York.

Zaldivar, O., J. Jose, and J. Otter. 1967. Phencyclidine for capture of stray dogs. J. Am. Vet. Med. Assoc. 150:772–776.

Zarrow, M. X., and K. Brown-Grant. 1964. Inhibition of ovulation in the gonadotrophin-treated immature rat by chlorpromazine. J. Endocrinol. 30:87–95.

Zimbardo, P. G., and H. Barry. 1958. Effects of caffeine and chlorpromazine on the sexual behavior of male rats. Science 127:84–85.

HAROLD MARKOWITZ

Oregon Zoological Research Center
Portland

Analysis and Control of Behavior in the Zoo

INTRODUCTION

The applicability of basic psychological techniques to zoo exhibits, design, and management has been the topic of much recent discussion. A number of pioneers in applied behavioral science, including especially Keller and Marion Breland (Breland and Breland, 1966) have argued that, while it is not possible to make zoos "natural," modern technology does provide capability for bringing forth from the animal more naturalistic behavior. For example, a limited budget will not support a forest for gibbons with appropriate climatic conditions; however, the first of the major behaviorally engineered exhibits, Oregon Zoological Research Center, does demonstrate that it is possible to encourage brachiation and "flying" around the cage while simultaneously providing the animals some entertainment and control over their own feeding.

AUTOMATED EXHIBIT DESIGN AND MANAGEMENT

The gibbon (*Hylobates lar*) colony at the Portland Zoo included two mature males, one female, and an infant male, when the research began, Although they were in a relatively spacious exhibit, observations of these subjects indicated that only one animal (a male) was exercising very extensively. This exhibit represented an unusual challenge because it was a mixed group that had never been removed from its home cage, and no exact precedent could be found in the literature. During pre-experimental

---

Preparation of this paper was supported in part by NIH Grant RR-0379.

observation there was some sharing of food between these animals, but no significant dominance hierarchy was apparent.

Two stimulus arrays separated by 30 feet were installed approximately half way (15 feet) up the back wall of the gibbon cage. The left panel included a large stimulus light, a buzzer, and a manipulandum consisting of a rod (projecting approximately 14 inches out from the panel) that activated a microswitch. The right-hand panel included an identical light and lever plus a food slot. Apples, bananas, oranges, and omnivore bread were cut in eighths and delivered by a belt constructed from 35-millimeter film arranged in a 54-foot endless loop. For the first 9 months electro-mechanical equipment constructed from surplus parts was used for programming. The remainder of the work was accomplished with solid-state programming apparatus designed in our lab. Cumulated data was collected with electromechanical counters and, for the last year, a printout counter was programmed to display data in 2-minute intervals.

During conditioned response (shaping) a remote control was used so that the experimenter could deliver reinforcers for successive approximations to the required response while moving freely outside the gibbon exhibit. This procedure was essential to minimize animals' responding to the experimenter's presence. Shaping was first accomplished to the right-hand apparatus and the subjects were reinforced for responses to this single manipulandum in the presence of the proximal light stimulus for 2 weeks. Although shaping was continued until each animal had made the required response a great number of times, there was some unevenness in number of responses per animal because some gibbons were more aggressive. Compromises were necessary to preclude extinguishing the responses of those that had learned at a faster rate.

After response to a single manipulandum was established, animals were once again shaped by successive approximation: first to move toward the left panel before responding to the right stimulus, and finally to pull the left manipulandum to initiate the chained sequence. A compound stimulus (light plus buzzer) was used for the initiation of each trial with the buzz lasting approximately 2 seconds. For a period of 5 weeks subjects were run with no time out between trials, i.e., the buzzer and left light were turned on immediately after food delivery. In the eighth week a 2-minute intertrial interval was used. After several months the public was invited to initiate trials within the 2-minute interval by depositing coins into a box in the viewing area. Considerable care was taken in the design of the accompanying instructions.

Research Contribution:

10¢ will start a trial when the light on this box is lit. The counter shows the total number of pieces of food earned by the gibbons today.

Animals are not machines and the gibbons may choose not to respond when the light is turned on. All money collected here will be used to develop more activities for our animals.

The apparatus was continuously available to the gibbons from 10 a.m. to 10 p.m. No other deprivation was involved and these animals had access to food for a greater portion of the day than the rest of the primate collection. At night lettuce and celery were provided *ad libitum*. Figure 1 illustrates a complete sequence of response.

Now that this exhibit has been in operation for nearly 2 years, it is possible to report a number of results. First, with respect to animal health, all of the behavioral signs, coat condition, and general health examinations have suggested that the regimen has been quite beneficial for the animals. With the exception of greens, the gibbons earn all of their food voluntarily. There have been no serious or prolonged illnesses, no scrapping to the point of tissue damage, and there has been much increased general activity. Second, the effect upon the public has been more universally rewarding than we could have anticipated. In general, people have been excited about making a contribution in this form. The $3,000 in dimes that this apparatus will collect this year will significantly help our research program.

It is interesting to note that during the transition from electromechanical to solid-state equipment, there was a considerable period when the apparatus was inoperative. Since the animals had had the freedom to feed any time of day, food was made available to them at all times when the apparatus was down. Observations conducted during these periods consistently yielded the same results. The gibbons would feed *ad libitum* from time to time but spent a considerable amount of their time (even when holding food in one hand) trying to turn the apparatus on. This included mounting and hugging the lights and apparently imploring the apparatus to respond.

Not only has the gibbon project been an interesting and profit-producing exhibit, but it has also generated volumes of data and produced the usual number of experimental questions. Prominent among these questions is the extent to which observed behavior is truly altruistic or cooperative and the extent to which one animal is simply using another. For example, without being gratuitously anthropomorphic, at present we cannot talk about how reluctant a gibbon holding a piece of food is to having it broken in half and shared with another. This and similar questions provided part of the rationale for a second major related project involving diana monkeys (*Cercopithecus diana*).

A family group—male, female, juvenile female, and infant male— were used in this case, working once again in the home cage. First the

Figure 1
Complete chain of response with automated gibbon apparatus: (a) gibbon operating the first manipulandum; (b) "flying" through air; (c) "brachiation"; and (d) operating the "pay-off" manipulandum where reinforcement is delivered.

subjects were taught to exchange large "Texas poker chips" for food by depositing the chips in a coin slot. After the value of the tokens had been established, the chips could then be used as easily quantifiable reinforcers. In order to obtain the tokens, the dianas (*C. diana*) had to accomplish a sequence of behaviors similar to that described above for the gibbons. Hoarding of coins, sharing, "defense of wealth," and other similar behavior was observed. Figure 2 illustrates several portions of the behavioral sequence involved in securing food.

Although this experimentation is still at an early stage, several interesting, reliable observations have already been made. The adult male dominates the scene, threatening the younger animals when they attempt to

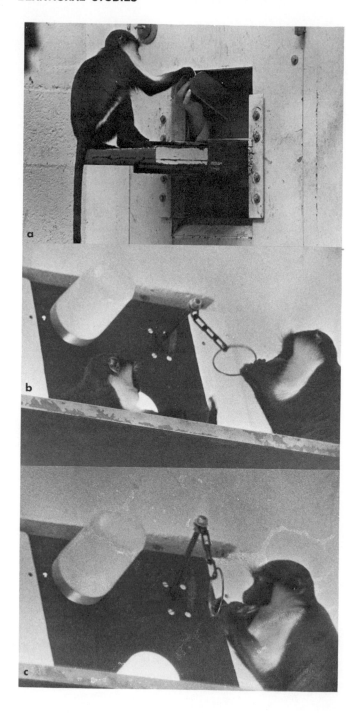

Figure 2
Diana monkey behavior with tokens in reinforcement project:
(a) earlier stages of "shaping" diana to deposit token in slot;
(b) diana pulling chain to obtain token; (c) diana releasing
chain, accumulating tokens in mouth; (d) diana depositing token
into slot; and (e) dianas awaiting the fruit reinforcer as a result
of token drop.

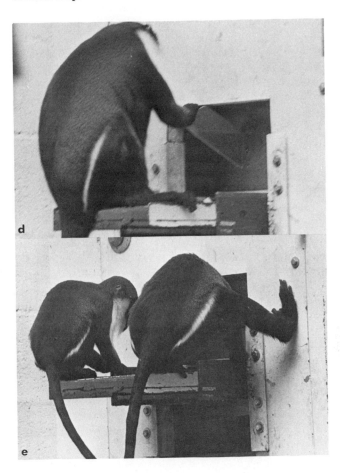

spend or obtain coins. However, he quite regularly feeds his mate and will
often deposit two coins, one right after the other, allowing the female to
have one of the pieces of food. Although supplementary feeding is continued to be certain that none of the younger animals is deprived,
"softening" of the male's behavior is evident as he becomes more and

more adept at "breadwinning," and he will occasionally let the youngsters have food during the experimental session. In the near future this exhibit will be interfaced with the public by means of a coin box much like that of the gibbons and, in the meantime, it has produced a regular clientele of repeat visitors who come to observe the diana progress.

Several additional automated exhibits to enhance zoo visitor participation are in various stages of development. These range from machines for involving the public more actively in learning about species, to a "tic-tac-toe" apparatus pitting great apes against human opponents.

The positive responses received from the news media, society supporters, and zoo visitors in general is a good indication that this kind of behavioral engineering program can work to improve zoo exhibits.

COMPARATIVE LEARNING

The best example to date of the research center's programmatic approach to a truly comparative picture of learning and problem solving is the design of an apparatus to test light–dark reversal discriminations with a wide variety of animals. Such an apparatus has been constructed for gibbon (*H. lar*), harbor seal (*Phoca vitulina*), elephant (*Elephas maximus*), camel (*Camelus dromedarius*), giraffe (*Giraffa camelopardalis*), and chimpanzee (*Pan troglodytes*). The following two summaries are illustrative of results to date as we begin addressing comparative learning questions.

In the case of the gibbons (*H. lar*), an unusual situation exists because, once again, the animals are not arranged in any single or individual organism design, but instead both a 4-year-old male and a 4-year-old female perform on display in the regular home cage. An apparatus with transluminated press panels has been installed within that cage; and a standard light–dark reversal discrimination procedure has been initiated with food delivered automatically for correct responses. To review quickly the mechanics of this procedure: The side where the light appears is randomized and the animal's first problem is to select that place. After this has been accomplished to a criterion of 20 successive correct responses, the situation is reversed and dark becomes "correct" with side still randomized. When an individual organism is run on this sort of task, the initial problem may not take very long—as few as a hundred trials—to learn always to go to the light. But, if the subject is a previously naïve animal, it takes an enormous number of trials (sometimes in the order of millions) to reach the first criterion on reversal. Having reached this criterion, as reversal implies, once again light is "correct" and, depending upon the particular animal and depending upon the species, there comes some point with these successive reversals where there begins to be some benefit—that is, a reduced number of trials instead of an accelerating number of trials to criterion (Markowitz and Becker, 1969).

With the two gibbons working together, the male learned the light correct problem very quickly (making only one error before reaching criterion), and the female took somewhat longer. However, when reversed, the male tended *not at all* to learn dark correct and, after considerable contemplation, we decided to run a cage criterion (i.e., criterion would be met as long as the total responses for that cage accumulated to 20 in a row, regardless of the animal doing the pressing). For several months we observed that our male, Milo, made light correct responses with virtually no errors, while Venus accelerated on dark correct. When Venus began making a series of dark correct responses, Milo would impatiently intrude himself and make an error (responding as if it were light correct) and she had to start from zero to reach criterion once again. The persistence of this "specialization" (through 197 reversals to date) was unexpected and provided an interesting avenue for analysis of differences between group and individual performance.

In the case of the harbor seals (*P. vitulina*), because of the logistics involved, we felt that it might be better to run an individual organism design. The animals haul out onto an island where only one seal at a time is admitted to the apparatus. The procedure is the same in terms of response requirements, but—instead of rewarding the animal with fruit—smelt or herring are delivered. Although we have only progressed to a maximum of five reversals, there is an interesting indication that harbor seals are quite adept at this learning task. Trials to criterion have consistently diminished (in one case from 2512 to 249 through five reversals).

Ultimately the Oregon Zoological Research Center plans to run a very large number of species on similar learning tasks (in the case of camels and giraffes, this may be the first basic learning research ever accomplished) to provide increasing data to address comparative questions.

One addendum concerning light–dark reversal discrimination: With some species the question in comparability of learning ability has always been measurement, or "How difficult is the response for that animal?" If you make a lever that is appropriate for a rat, it may not be easy for a pigeon to press; a key that withstands a hippo's destructive powers may not be operable by a mouse, and so forth. What we have tried doing in any situation where this question has arisen is to use capacitance detectors rather than switch-loaded panels. Large stainless steel response panels serve as antennae, and the animal need only come into contact with them to register a response. This allows the animal to select response modes over a wide range. The detector is tunable so that subjects from shrew to elephant can operate the same apparatus by coming into contact with it. The giraffe, for example, selects to respond by licking the apparatus.

The camel surprised us somewhat. Prior to reversal testing, Dr. Vic Stevens thought it might be interesting to see how the camel would respond to high ratio schedules. At what point would "strain" begin to show? (i.e., at what point would the camel no longer be willing to work to receive one reinforcer for "X" number of presses?) We expected to see the camel, in his first bit of operant learning ever recorded, come up to the panel and "nose" it. What we were not prepared to observe was this animal's unique solution to the effort requirements of operant work. The camel walked into the appropriate area, put his face up against the panel, and (apparently in order to maintain the folklore about camel idiosyncracy) responded by vibrating his chin at very high frequencies. This enabled him to make 150 responses for one reinforcer without any significant sign of strain.

The reversal discrimination procedure, in addition to the fundamental cross-species comparisons, is beginning to yield adequate pretraining for other kinds of discrimination testing and sensory assessment. For example, from the elephant's learned discrimination abilities may come an assessment of visual acuity in this species in the near future.

EFFECTS ON INDIVIDUAL AND GROUP BEHAVIOR

A second major research question involves data concerning long-term effects of mastery of the instrumental learning procedures. Since it is our belief that behavioral engineering can provide sound partial solutions to animal boredom and inactivity, careful assessment of general effects is a critical part of the continuing program at the research center. Much of what follows is prospect and an earnest request for help in development of "norms" for animal health, but first some initial approaches to these questions.

There are a large number of student interns involved in research at the Oregon Zoological Research Center and many of them have elected projects relating to the effects of unique feeding regimen on behavior. Erni Soper's recent master's thesis (1973) discusses in detail the effects of changed feeding procedures on the center's diana monkey group. He suggests that there is increased focal aggression when food is fed several times a day as opposed to being delivered all at once. This is an important observation that deserves considerable additional experimental attention. Two other coworkers, Tom Cox and Gene Stainbrook observed effects of feeding drills (*Mandrillus leucophaeus*) small amounts of food at various daily intervals for several months, and their preliminary data show some interesting trends. Aggression increased initially and the juvenile in this situation (adult male, adult female, juvenile male) was not allowed near the food. But, as the new procedure became increasingly familiar to the drills, aggression rapidly approached pre-experimental levels and the young drill was allowed free access to food.

Preliminary results with several other species suggest the general hypothesis that, for many animals, new requirements may temporarily focus some aggression, which disappears as the subjects become more adept. It should also be emphasized that none of this temporary food-associated aggression has resulted in injury to animals in the newly engineered exhibits. It appears quite similar in form to brief intraspecies food competition in the wild with conditions of relative abundance.

The Portland Zoo was recently provided an unusual opportunity to study the long-term retention of operant training. Eight years ago Dr. Leslie Squier tested a number of elephants (*Elephas maximus*) at the zoo and, because of some technical problems, had to abandon his plans before their completion. However, by fortunate coincidence, he had completed training three adult females to perform a light–dark discrimination task very similar to the one described earlier in this paper. His old apparatus was literally on a scrap heap, but we were able to refurbish it so that it assumed close to its original form. With Dr. Squier's cooperation, we embarked on a test of memory with an 8-year intertrial interval. Much to our surprise the first subject (Tuy Hoa, b. 1955) immediately approached the apparatus and began to respond (usually, it takes weeks to train a naïve elephant to begin to reliably respond to an operant panel for sugar cubes). She made five straight correct light responses followed by one error and then proceeded to criterion without further mistakes. It took Tuy Hoa 6 minutes to demonstrate her retained learning in a situation to which she had not been exposed for more than 8 years.

Rosy (b. 1949) and Belle (b. 1952) produced markedly different results. Although both were willing workers, it took 1,240 trials for Rosy to reach criterion, and Belle required considerable "shaping" with a total of 3,199 trials to obtain 20 consecutive correct. During these long testing sessions (2 hours per day) it appeared that these two elephants were having some significant visual problems and, as a result of experimentation, we discovered important physiological problems in the visual system of Rosy and Belle. In both cases there was atrophy of vascularization in the retina as compared with Tuy Hoa. This is apparent in direct ophthalmologic examination and in slides that have been produced from photographs of the retinae. Discovery of these visual deficiencies in our elephants illustrates one of the fringe benefits from active zoo research:

Regular testing of the animals serves to focus sufficient attention on behavior to identify gross abnormalities that might otherwise go unnoticed.

In this same vein we are currently working on a proposal with Dr. Michael Schmidt, Portland Zoo's veterinarian, to establish a series of medical criteria for carefully evaluating the effects of our new behavioral regimen on general health parameters. It is our mutual hope that these efforts will contribute to providing active, preventive health care for zoo animals. We are continually soliciting help in developing health evaluation standards; unfortunately, however, in many cases the only available criteria are based on absence of *severe* illness.

Other projects of the research center include work toward the telemetry of bioparameters as briefly described to AAZPA last year (Markowitz, 1972). These techniques will enable on-line assessments of the effects of instrumental conditioning on such variables as temperature and heart rate. The addition of Dr. Schmidt to our staff has provided much needed medical expertise in this area.

COOPERATIVE BEHAVIOR

Attempts to quantify cooperation among the diana monkeys were described earlier. These same gibbon studies have also generated some data on cooperation, despite difficulties in deciding at what point there is true cooperation and at what point stealing of food begins. We have observed on many occasions (Markowitz, 1973) that one animal will work on the remote lever while another operates the pay-off manipulandum. This is especially apparent with the female, whom the two male animals appear quite willing to feed. In contrast, these males are more reluctant to feed each other by cooperatively performing the response requirements.

There are a number of other designs that much more clearly require cooperative behavior, including one used in the center's laboratory involving a transparent partition down the center of some chamber or living situation. When the animal on one side of the partition responds, food is delivered to the subject on the other side of the divider; after some period of time, the situation is reversed. This "you scratch my back, I'll scratch yours" design has produced very mixed results in the laboratory, and for a number of aesthetic reasons it seems inappropriate for the zoo.

In the general realm of cooperation, Katheryne Johnson is conducting a study that involves the teaching of American Sign Language to young chimpanzees (*P. troglodytes*), ranging in age from 1½ to 3½ years. Although the vocabularies are not yet very extensive, at this point there is an interesting chaining of words to make partial sentences. Our oldest chimp knows 13 signs with 100 percent accuracy and 6 signs with 75 percent accuracy. We have seen considerable observational learning in this situation. The very first sign learned by the 2½-year-old female was copied from the 3½-year-old male. As this project progresses, one focus will be the extent to which there is intraspecies communication. Although there have been similar major research programs in this area (Gardner and Gardner, 1969), data on intraspecies communication are quite limited.

In conjunction with this project we have developed a teaching machine to aid in the acquisition of vocabulary and the connection of nouns and verbs for our chimps. This machine has two sets of press panel "windows," the top set arranged in a match-to-sample configuration and the bottom pair randomized. A picture (e.g., a photograph of an apple) is displayed in the top center window, and the animal's task is to match by pairing the appropriate American Sign Language hand symbol, which appears on one of the side windows. Correct matching turns on the bottom display where the sign for "eat" and "drink" appear. Thus, correct sequential selection of "apple" and "eat" delivers a piece of apple; "juice" and "drink" produces juice, etc. Our advisors, who were teachers of the deaf, repeatedly suggested that this device would be more useful with auditorily handicapped children than many of the available programs. One of our recent grant proposals to develop a machine for deaf children may be identified as another direct product of animal experimentation.

CONCLUSION

Throughout this article I have tended to use plural nouns, not to be fatuous, but because all of the work described could not possibly have been accomplished without hundreds of people helping to direct and run the research. These coworkers have demonstrated that it is possible to create more interesting environments for animals without excessive cost (even to economize in areas such as food usage). Replication of similar exhibits has been greatly simplified and reduced in cost as a function of new developments in micro-electronics. Cost is no longer an adequate excuse for failure to begin re-engineering zoos to allow animals more active exhibit components. There is a responsibility to do what we can to improve the lot of the animals we have brought into captivity. This responsibility extends beyond breeding and observation programs, although these are essential. We *can* improve the lot of the animal by making his life less boring, providing him some control of his environment, and encouraging him to be more active.

REFERENCES

Baldwin, J., V. Stevens, and H. Markowitz. 1973. Operant conditioning for ungulates in a zoo setting. Ankus (Portland Zoological Society) 5:17–19.

Breland, K., and M. Breland. 1966. Animal behavior. Macmillan, New York.

Gardner, B. T., and R. A. Gardner. 1969. Two-way communication with an infant chimpanzee. *In* A. M. Schrier and F. Stolinitz, eds. Behavior of nonhuman primates, Vol. 4. Academic Press, New York.

Markowitz, H., and C. J. Becker. 1969. Superiority of "maze-dull" animals on visual tasks in an automated maze. Psychonomic Sci. 17(5):171–172.

Markowitz, H. 1972. Research in the zoo. Proc. 48th Ann. Am. Assoc. Zool. Parks Aquariums Conf.: 99–102.

Markowitz, H. 1973. Behavioral research in the zoo. Ankus (Portland Zoological Society) 5:6–11.

Soper, E. 1973. A master's thesis on effects of changed feeding procedures on the center Diana monkey group. (Unpublished.)

BENJAMIN B. BECK

Chicago Zoological Park
Brookfield, Illinois

Student Behavioral Research in Zoos

Many zoos place research and education among their primary functions. However, limited resources often prohibit full realization of the potential of zoos in these areas. Student research, which places minimal demands on staff time and budget, frequently results in very respectable scientific data while providing a powerful educational experience. This combination suggests that more consideration should be given to programs of student research in zoos.

I first became personally aware of the unique advantages of student research when, as a college instructor, I led 13 undergraduate students on a month-long field study of the ecology and behavior of animals using a waterhole in Arizona's Sonoran Desert. The students participated in all phases of the design and execution of the study. Only one of the students has since specialized professionally in biology, but all learned methods in inquiry and attitudes toward natural resources that serve them well as educated citizens. Additionally, some of our data have been published in a scientific journal (Beck *et al.,* 1973). This experience taught me that students learn more as active discoverers of knowledge than as passive recipients of knowledge. The study cost less than $6 per student per day. We were able to have this productive experience at such a low cost because we worked at a national monument where interesting animals and habitats were already being maintained for public recreation and education. This is precisely the factor that makes zoos such attractive sites for student research.

One might well ask whether I advocate that zoos throw open their gates to eager students simply for the intangible benefits derived from noble endeavor. I feel that, on the contrary, there are concrete advantages that accrue to zoos that host student research. First, the educational programs of most zoos are targeted mainly to the general public whose visits to the zoo are infrequent and of short duration or to groups of students from local schools. Without questioning the effectiveness of these important programs, it is fair to say that they do not meet the needs of students with an unusually strong interest, aptitude, and/or background in zoology. By providing such people with the opportunity to conduct research, zoos can add a vital dimension to their educational programs. Second, because researchers devote considerable time to studying only a small portion of an animal collection, they can pick up subtleties about their subjects that escape a general curator or supervisory keeper. These details can provide feedback for improved husbandry, more effective design, and ideas for greater educational impact of the exhibit. For example, students of an animal behavior class from a local college have been observing Brookfield's herd of Père David's deer. The reproductive performance of the herd had been disappointing in that only one of six sexually mature females had produced young. The dominant stag had been seen to copulate with only this female, ignoring the other five. We were puzzled as to why a second, subordinate stag did not copulate with the others. The student researchers provided hard data showing that the dominant stag actively herded all females and subtly kept the subordinate male from them. With this information we cut out the dominant, and the subordinate has now been seen to copulate with several females.

Another advantage of hosting student research is to form a reservoir of talented people with knowledge of and experience in zoo operation. Zoos will need such people to meet successfully the challenges they now face.

A fourth benefit derives from the inherent and compelling interest generated by research, particularly by research in animal behavior. Trustees, zoological society members, and the professional community become increasingly involved and committed, thus enhancing the zoo's base of support.

A spectrum of possible relationships between zoos and student researchers follows; some or all of these relationships have undoubtedly been used at other zoos, but—as there are few published reports of such arrangements per se—I'll draw on our own experiences at Brookfield. I have borrowed from Rumbaugh's descriptions (1971, 1972) of his relationship with the San Diego Zoo for student research and teaching. It is heartening to note the similarity between his experiences and recommendations from the perspective of an instructor utilizing a zoo and my own from the perspective of a zoo staff member. I should note that it is general policy at Brookfield that research is conducted on animals in the collection rather than in research colonies not open to the public. Further, research is not allowed to conflict with or detract from the general exhibition program. Behavioral research is ideally suited to such a policy, and most of our work has therefore revolved around such topics as social organization, infant development, and social learning where little intervention or manipulation is necessary.

High school and college students periodically volunteer to serve as assistants to the resident scientific staff on a volunteer basis. Most seek such a position during the summer, but others capitalize on the flexible programs of some area schools to receive academic credit for working for longer periods during the school year. At Brookfield we get about six requests a year from students to serve as volunteer research assistants. We accept most, principally on the staff's subjective impression of the sincerity and motivation of the student. We do not demand extensive academic background in the biological sciences. Upon acceptance, the student is asked to write a brief account of his planned work for in-house communication.

Our experience with student research assistants has been most satisfactory. This success is due, I think, to viewing these students as junior colleagues. After a short period of training, they conduct first-hand observation. Usually they study a limited aspect of the behavior of only a small group of animals, thereby enjoying intense involvement, which is strongly motivating. They are rarely asked to do uninteresting, routine chores such as filing and, in fact, are encouraged to design and conduct original projects commensurate with their increasing skill and experience. This freedom and responsibility intensifies their motivation and generates strong interest in analyzing the data they have gathered and compiling it for publication. Several are co-authoring scientific papers with the resident staff members with whom they worked. These students acquire surprising competence in understanding animal behavior that is primary, practical, and specialized; they do not get the broadly based theoretical orientation that is more easily acquired in the classroom. They learn the basic skills of inquiry, which are applicable to any field of endeavor, and gain a feeling of mastery from very tangible achievement.

If funds are available, we sometimes offer a modest salary to those student volunteers who become deeply involved in their projects for long periods of time. Remuneration neither detracts from nor adds to the performance of those students who have already proven their conscientiousness and capability as volunteers. A volunteer research assistant requires that a permanent staff member, actively engaged in or at least interested in zoological research, serve as a supervisor. From a financial standpoint, the student's assistance compensates for the time the staff member devotes to the student, but supervision of more than two students at a time may compete with the staff member's other duties and thus has to be debited, at least theoretically. The equipment and supplies used by the students need not exceed those already budgeted for the staff member. As noted, salary is not necessary but the student should be covered by accident insurance. In all, students serving as research assistants enjoy very productive educational and research experiences at a low cost to the host zoo.

A second mode of student research is that in which an advanced undergraduate or graduate student conducts a circumscribed research project of his own design. Behavioral observations comprising substantial parts of both master's level (Susman, In press) and doctoral (Avis, 1962) research on primate locomotion were conducted at Brookfield. Researchers on these projects spent more than three months each at the zoo. In contrast, supplementary observations for graduate student research on primate dietary specializations (Walker and Murray, 1973) were conducted in a single morning. We receive about 12 inquiries annually from advanced students wanting to conduct original research. We have found it necessary to prepare a standardized response to such inquiries:

The following policy guidelines were formulated in discussion by the research staff. They are intended to help us deal equitably and ethically with requests for research assistance from advanced students while protecting the legitimate interests of the zoo staff and the zoo in general. A secondary purpose is to help us document aid rendered to other institutions and people.

Any proposed new research project should be brought to the attention of the Associate Director, Research and Education, who will confer with the Associate Director, Animal Collection, and others as appropriate. As soon as practicable, students should outline their proposed research *in writing.*

We will issue a complimentary pass and letter of introduction for students whom we agree to help. *Per se,* these documents assure the student of no special privileges beyond free admission and parking and general courtesies from employees. To facilitate student studies, one of the zoo research staff will ordinarily act as a principal zoo contact person or sponsor for each student. Unless special arrangements are made, the zoo sponsor will not substitute for the student's faculty advisor for evaluation purposes.

As with established outside investigators, student research projects will be considered only to the extent that they are compatible with the welfare of the animal collection, the ongoing research of the staff, and prior commitments to other researchers. Arrangements for manipulation of animals, as by feeding, shifting, marking, etc., should ordinarily be worked out in advance with the curators and their keepers.

Students will be expected to act as consultants to the zoo staff on the management of the animals they study and on educational exhibit ideas that may be generated by their research.

In the event of any publication or edited film resulting from the research conducted at the zoo, the student is expected to make a general acknowledgement of the Chicago Zoological Society (for provision of study site and materials and privileges). A copy of the publication or film should be deposited in the zoo library. Use of animal collection data, such as animal histories or other curatorial material may properly be attributed to personal communication from the curator or other staff member most directly involved. Use of photographs or tape recordings or similar material should be reflected in a general credit or a personal acknowledgement, if the staff member involved is known. Use of other information from zoo staff must be arranged on an individual basis. Staff are under no obligation to supply data on behavioral profiles, group social structure, etc. If such material is given to a student, credit should be in whatever form is mutually agreeable, which may range from co-authorship to a footnote.

Several points in this policy statement deserve emphasis. First, we require students to submit a written proposal detailing the objectives and procedures of the project. This helps to separate sincere, well-formulated projects from those which are only casually conceived. It also gives us a concrete basis for judging the value and feasibility of the project and, when approved, serves as a "contract" stating in advance the needs and expectations of both the zoo and the student. For these reasons I cannot overemphasize the importance of requiring a written proposal. If an advanced student cannot state the goals and procedures of his project in writing, he probably does not know what they really are. If he does not know what they are, his project may founder or he will make unanticipated demands as the project develops. If these demands cannot be met by the zoo, frustration and hard feelings will result and the relationship becomes counterproductive. It is perfectly reasonable and professionally proper to require a written statement of intent and needs, in advance, from a visiting researcher and for the sponsoring zoo to respond with a written invitation and statement of terms of cooperation. While awkward and formalistic, we have found this exchange to be invaluable.

Another important point of the policy statement quoted above is that we make explicit our expectation that the student serve as a consultant to the zoo on the animals and exhibits he studies and that he make available to the zoo all data and other materials that result from his research. The student researcher is thus informed that it is his personal and professional responsibility to include improved zoo management as a goal of his work.

Visiting student researchers are unsalaried, but they should be covered by accident insurance. We sometimes provide such consumable supplies as film and such services as shop time for constructing simple equipment. We may allow use of such zoo equipment as cameras, binoculars, stopwatches, and library materials. These costs must be reckoned but, as they are agreed upon in advance, they need not be excessive. The research undertaken by visiting student researchers is commonly sophisticated and intensive and thus may require considerable assistance from professional staff and keepers. This may involve shifting or marking animals, implementing special feeding regimes, providing animal records, photographing, supervising "after hours" research, and monitoring the close contact between researcher and animals, which is often necessary. These activities compete with the employees' routine duties and must be debited. We reject proposed research that requires such excessive assistance as to necessitate supplemental staff salaries. The costs of hosting visiting student researchers exceed those for student research assistants; however, these can be controlled in advance so as not to exceed available resources. Visiting students can conduct very productive original research requiring no more financial assistance than that available in even the most modest zoo budget. Because of their advanced training and sophistication, these students can be expected to produce results that gain greater circulation and attention in the professional community and are apt to be more useful as feedback for effective animal maintenance and exhibition.

A third vehicle for student research at Brookfield is study conducted by an entire college class. This is not to be confused with brief survey visits but rather consists of students making first-hand observations for periods ranging from a day to several months. The most common goals of the short duration studies are to acquaint students with the techniques of behavioral study and with the difficulties and complexities involved in definition and measurement of behavioral phenomena. Several students may watch the same animal group for a few hours and then compare notes to learn about interobserver reliability. They may score only one behavior (my favorite is "presenting" by monkeys) to learn that there are many variations within what is often simplistically presented as a single behavioral category. They can be asked to measure "dominance" in a social group to become aware that this phenomenon, used so glibly in texts, is extraordinarily complex. Students can learn more about the concepts and methods of behavioral study in a few hours of primary observation than they can learn in a semester of lectures and readings.

Some instructors actually use Brookfield as a classroom and have the students conduct research for the entire term. In one case, each class member studied a different group for about 20 hours per week. Not only did the students collect good data on the studied groups, but their final papers showed considerable familiarity with the literature on that species and competence in observational techniques. Periodic seminar discussions allowed interspecies comparisons and provided a broader, theoretical perspective. Another class studied only one group, the Père David's deer mentioned above. By assuming shifts, they collected an enormous amount of data in a short time and conducted an exhaustive literature search. Their intensive group effort produced a communal or cooperative atmosphere and the shared responsibility was strongly motivating. I should note that none of these students had any prior experience in animal behavior. "But they're not doing anything" was the group's response when we stopped in front of the deer paddock in their initial search for a suitable study group. It was a hot day and the deer were immobile in their wallow. I mentioned thermoregulation and seasonal and daily activity cycles, and they grudgingly agreed to watch a while. Several hours later they were bubbling with questions, answers, theories, and plans. In a matter of months this naïveté metamorphosed into surprising sophistication.

Aside from the staff time involved in getting these projects started, research by students in college courses has cost the zoo nothing. The students, of course, expect no salary, their instructor/supervisor is paid by their college, and their equipment is provided by their college. Since they are beginners, their research always consists simply of observations that can be conducted from the public space. Research by college classes is a most inexpensive and productive enterprise and is thus ideally suited to any zoo.

In 1973, we initiated a new program at Brookfield, a fourth mode of student research, in which graduate and undergraduate students were invited to design and conduct behavioral research for three months during the summer. The program was conceived in response to increasingly heavy demand from faculty and students for more student research opportunities at the zoo. Members of our Board of Trustees, recognizing the value of student research, personally donated $9,000 for student stipends and expenses.

The program was announced in advertisements and mailings targeted to colleges and universities having faculty members actively engaged in behavioral research. Students were invited to propose a behavioral study that could be conducted in the zoo within three months. The proposal had to state the general and specific aims of the project, detailing procedure and demonstrating familiarity with pertinent literature. Applicants were also asked to submit academic transcripts and letters of recommendation. We received 36 completed applications, which were reviewed by the zoo's professional staff and members of the Board of Trustees. We looked primarily for projects that were sound, compelling, and compatible with other zoo functions. The student's credentials were reviewed simply to ensure that he was academically prepared to conduct the research he proposed and that he had the ability to work independently. Seven students were chosen, with surprisingly little disagreement among the reviewers, to participate in the program. A freshman from the College of

DuPage conducted a study of the behavior and social organization of our Dall sheep herd. A senior from the University of Hawaii concentrated on the same phenomena in the Siberian ibex herd. A third-year graduate student from the University of Wisconsin (Madison) studied the locomotion of two species of galagos. Also studying primates was a junior from the University of California (Davis) who worked on vocalizations and activity cycles of lar gibbons, and a second-year graduate student from The Johns Hopkins University who did a fine-grain analysis of play in a juvenile group of hamadryas baboons. A Yale sophomore studied social behavior and spacing of our rock hyrax colony, and a first-year graduate student from the University of Illinois (Urbana) compared the facial expressions of some Old and New World monkeys. Five of the seven participants were women and four of seven worked with primates; we cannot explain these imbalances, although the latter may be due to the current popularity of primate research.

Each student was assigned a professional staff member as advisor, although we emphasized that the student himself was solely responsible for design and execution of the project. Staff and program participants met as a group once a week to discuss problems, emerging data, and general scholarly issues. Two distinguished figures in the field of animal behavior led special seminars. The students were expected to analyze their data and write papers suitable for publication before leaving.

With a few, inconsequential exceptions, the program was eminently successful. The atmosphere was electrified with discovery and learning. As predicted, the student's authorship of and responsibility for the project was intensely motivating. Each averaged at least 60 working hours per week and could commonly be found in the zoo nights and weekends. There was great mutual support and cooperation among and between students and staff. Six students have submitted their papers; the seventh is completing the last of several self-imposed revisions. All of the papers contain data that genuinely contribute to the field and at least several of the papers are publishable as they now stand. Most of the students provided explicit and valuable suggestions for improving the husbandry of their subject animals and for the effectiveness of the exhibit. The trustees have voted to repeat the program in the summer of 1974. We all profited greatly and enjoyed the exhilaration that comes from accomplishment and discovery.

Each student was awarded a stipend of $1,000 and was covered by accident insurance. Each was responsible for his own transportation and living expenses. Required consumable supplies, such as film and recording tape, were provided. Small honoraria were paid to the guest seminar leaders. These costs and the stipends totalled about $8,700. Additionally, the scientific staff, the zoo photographer, two secretaries, many keepers, and other zoo employees allotted considerable time to the program. While these services did not entail supplemental salaries, they must be debited, although an exact figure is difficult to calculate. The use of equipment, office space, and other zoo facilities must also be considered but, again, no supplemental monies were dispensed. If one estimates a grand total of $2,000 per student, surely a high figure, this is still quite reasonable for a very productive 12-week program and within reach of many zoo budgets. The American Association of Zoological Parks and Aquariums might profitably consider seeking and designating funds for stipends and overhead costs for the sponsorship of a national summer student research program. An AAZPA committee could receive and review applicants and match worthy proposals to the staffs and animal collections of cooperating zoos.

I should mention some of the problems we have encountered in our experiences with student behavioral researchers. None is unique to the presence of students; all are apt to be encountered in the conduct of scientific research in a zoo by anyone.

The most important problem is that scientific research, even behavioral research, can interfere with other functions and in the day-to-day operation of the zoo. It may be necessary to mark the animals, which detracts from their appearance in a naturalistic exhibit. Temporary separation of individuals, which could weaken the integrity of social groups, may be desirable. It may be useful to manipulate feeding schedules, which could temporarily irritate animals and disrupt keeper routines. The list is endless, but all decisions concerning research intervention should be dominated by the principle that illness, injury, prolonged psychological stress, or surgical insult should be avoided. The most important zoo asset is its animal collection, and research cannot detract from the health or well-being of any zoo animal. Research that causes minor discomfort or disturbance of short duration may be appropriate but must be carefully considered. Likewise, temporary detraction from the esthetics of an exhibit or disruption of employee routines may be warranted and have to be balanced with the potential worth of the research to the zoo itself and to the scientific community. Written proposals allow advance calculation of the nature and degree of interference and thus are indispensable in deciding whether or not to host a particular project. We have found that very valuable scientific research, especially behavioral research, can be conducted without consequential interference with other zoo functions.

Many projects are more appropriately executed in laboratory colonies or in the field; zoos should recognize this and sponsor only those projects that are compatible with their own goals and operations. Zoos must actively encourage research that is appropriate and make an effort to accommodate the minimal interference that may accompany such activity. Signs explaining the research to the public may actually convert the interference into an asset.

Another problem is that many zoo employees (particularly keepers) who devote their career to the efficient operation of the zoo for public recreation and education and for conservation, are suspicious of arrogant intellectuals who arrive suddenly to conduct esoteric scientific studies of a relatively short duration. Zoo administrators must take pains to point out to all that scientific research (which does not conflict with other zoo functions) is a legitimate and important aspect of the zoo operation. They must introduce researchers to zoo employees, particularly those most directly involved, and make sure that the project's aims, procedures, and potential benefits to husbandry and exhibition are carefully and understandably explained. The source of the project's funding and its relationship to the total zoo budget should be clarified. The researcher, for his part, must recognize and respect the investment that the zoo employees have in their work. He must share their concern for the animals and the public and must accommodate his research to the daily routine of zoo operation. He must make sure to explicate the rationale and results of his work. In brief, this problem is one of arrogance and ignorance; the solution is respect and communication.

A third potential problem is that students, by definition, have not solidified their career goals and live in a cultural environment that is rapidly shifting. In some cases this is manifest in lightning changes in even short-term interests. Thus a student may express his intention of conducting a research project and may even begin work, and then suddenly withdraw. Again, we have found the written proposal to be the best solution to this problem. A student who invests the time and energy in formulating a proposal is apt to be firmly resolved and motivated. I estimate that about 70 percent of those students who inquire, usually with great enthusiasm, about research opportunities at Brookfield never submit a written proposal. On the other hand, I can think of only one student who submitted a written proposal and did not carry his project to completion. The written proposal is an excellent device to eliminate those with only fatuous interests.

Another problem that we have encountered is that many contemporary students present an appearance or life style that clashes with that of some employees and administrators. The bulk of this issue, of course, is a rapidly closing cultural gap, which generates friction in many spheres of contemporary life and demands only mutual respect and toleration for solution. Nonetheless, safety demands that feet must be shod and that long hair and dangling baubles be controlled (especially when working with primates). The same of course applies to neckties.

A related problem is that most students are quite naïve about the genuine dangers of injury and illness that exist when working in close proximity with seemingly benign wild animals. Some tend to be negligent with expensive and sophisticated equipment, which may be provided by the zoo to assist their research. Harm or damage from either of these sources is easily prevented by advance instruction.

A final problem is that college or high school faculty may shift the responsibility for academic supervision or evaluation of student researchers to zoo personnel. Staff members may be willing and/or qualified to accept these responsibilities, but there should be an understanding in advance. If these roles become routine, it is quite legitimate for zoo staff members to seek remuneration or academic appointments.

Zoos acquire and maintain large and varied collections of wild animals for the general public's education and recreation and for conservation. Many wild animals are, for quite legitimate reasons, increasingly difficult to obtain and most require sophisticated and costly maintenance and veterinary care. The numbers and kinds of wild animals kept in most laboratory colonies are therefore decreasing. At the same time scientific interest in these animals is growing rapidly. Zoological research, then, increasingly becomes confined to the field or to zoos. Field research is expensive, especially for students, and many problems cannot be studied in the field setting. Zoos must therefore recognize their responsibility to make their collections available to researchers when their projects are compatible with other zoo functions. Researchers must temper their demands for colonies of wild animals for their exclusive scientific use. The increased use of zoos for research is dictated simply by optimal use of rare resources.

The public of most countries remains committed to the support and use of zoos. It is imperative that zoos include and strengthen research within their functional sphere. To do otherwise is sheer negligence. I hope I have made a convincing case that hosting behavioral research by students is not an excessive or unjustified drain on zoo resources and presents no insurmountable conflicts with other zoo functions or with day-to-day zoo operation.

REFERENCES

Avis, V. 1962. Brachiation: The crucial issue for man's ancestry. Southwest. J. Anthrop. 18(2):119–148.

Beck, B. B., C. W. Engen, and P. W. Gelfand. 1973. Behavior and activity cycles of

Gambel's quail and raptorial birds at a Sonoran Desert waterhole. Condor 75: 466–470.

Rumbaugh, D. 1971. Zoos: Valuable adjuncts for the instruction of animal behavior. Bioscience 21(15):806–809.

Rumbaugh, D. 1972. Zoos: Valuable adjuncts for instruction and research in primate behavior. Bioscience 22(1):26–29.

Susman, R. In press. Facultative terrestrial locomotor hand postures in an orangutan (*Pongo pygmaeus*). Am. J. Phys. Anthrop.

Walker, P., and P. Murray. 1973. An assessment of diet and masticatory efficiency in a series of anthropoids. Paper presented at the Ninth International Congress of Anthropological and Ethnological Sciences, Chicago, Illinois.

GORDON M. BURGHARDT

University of Tennessee, Knoxville
and
Knoxville Zoological Park

Behavioral Research on Common Animals in Small Zoos

Ten years ago research in zoos would have attracted little or no interest; in fact, it would have been possible only in the larger and more famous zoos. While the role of research in zoos is still controversial, it is clear that *behavioral research of some kind is essential* to the survival of zoos. Small zoos must also decide how and if they will contribute to this survival, whether to be content to be parasites or to muddle through on a trial-and-error basis.

THE STUDY OF BEHAVIOR IN ZOOS AND ACADEME

The study of animal behavior is a respected topic of specialization, research, and teaching in zoology, psychology, and biology departments in colleges and universities throughout the country. Prior to 1960 most biologists considered the study of behavior unscientific or "soft," in spite of the tradition of European ethology and early American zoology. Comparative and learning psychologists have traditionally studied nonhuman animals, but by preference and habit, for economy, time, and efficiency, they mainly studied domestic animals under well-controlled, uniform,

This work was supported, in part, by Grants MH 15707 and MH 20565 from the Public Health Service. The Great Smoky Mountains National Park and the Maryville College Environmental Center provided essential facilities. I also thank Ellis Bacon and Cheryl Pruitt for permission to use unpublished photographs and data, Lori Burghardt, and Doris Carey.

and artificial laboratory environments. Furthermore, they focused on behaviors that could be easily measured and manipulated, such as bar pressing and maze running, in an effort to study such general topics as learning, hunger, and visual discrimination. Since the methods and apparatus used by psychologists did not lend themselves readily to use in zoos, zoos were not used.

Today the picture has changed radically. *Zoos need researchers and the animal behavior and comparative psychology researchers need zoos.* The question is whether the mutual and individual needs of zoo and research can be effectively meshed into a harmonious system in small institutions, in particular. I think this is not only possible but necessary; yet it is not inevitable. Success will necessitate some understanding by, and education of, both zoo personnel and animal behavior researchers in the goals, purposes, and techniques of the other. If such contacts are as important as I think they are, then they are too vital to be entrusted to a few major zoos. Smaller institutions, especially in areas close to colleges and universities, will increasingly be called on to cooperate and provide facilities for student and faculty research.

For years there was little contact between animal behavior research and zoos. Some outstanding exceptions included Hediger and his associates (e.g., Hediger, 1950), who had close contacts with European ethologists. Most of the noteworthy efforts of academics to use zoos centered on primates. Zuckerman (1932) constructed some controversial theories of primate behavior through observation at the London Zoo. About 1950 in the United States, primarily under the influence of Harry Harlow, laboratory animal psychologists became increasingly interested in primates, particularly the rhesus monkey. Harlow did some research on monkeys at the Vilas Park Zoo in Madison, Wisconsin, but this work was primarily a supplement to laboratory research utilizing animals housed in small metal cages. Zoos have not been completely ignored, therefore, by academic researchers nor did zoos completely neglect research. It is often desirous to have research personnel lodged within the zoo institution, but the problems and goals of in-house and out-house scientists are often similar.

WHY ZOOS NEED RESEARCH

The zoo movement itself has matured and become much more organized in the last few years. There is concern with cross-fertilizing, formalizing the exchange of information, and developing more professional personnel. The forces relating to research from the zoo side operate in three interrelated areas: breeding, caretaking, and exhibiting.

Breeding

Increasing concern that the stockpile of animals for zoos is rapidly declining should convince zoos that they must encourage breeding programs within their own and other zoos. This point has been extensively developed in earlier papers at this symposium and elsewhere. Suffice it to say that successful breeding programs should be extended to all species, not just those presently endangered, and exhaustive research work is needed.

Caretaking

As soon as one is not only interested in, but actually dependent on, the breeding and handrearing of wild animals, it becomes apparent that these rare and valuable animals cannot be entrusted to zoo keepers with low pay and the status of garbage men. Proper care of these animals entails sophisticated knowledge, dedication, understanding, and a certain amount of risk to personal safety. Under these circumstances, it is not surprising that the animal enthusiast who becomes the zoo keeper and is happy with minimal recognition and salary is all too rare. Persons with a love for animals, basic zoological knowledge, an observant eye, nose, and ear, a knack for asking relevant questions about an exhibit, for knowing the answer to other questions, and having a fairly detailed knowledge of the specific species under their care are going to be increasingly needed. Such individuals, or their absence, can have potent effects on research.

Exhibiting

A third problem is that zoos are confronted with the demand of the public for realistic and humane displays. Many people claim that zoos, as they have historically operated, are outmoded if not immoral. According to some critics even exhibiting animals in cages that are spacious by old standards but miniscule by natural standards is to be deplored and opposed. I encounter this view particularly among college students and educated people with an "ecology ethic." We cannot afford to dodge the views of an often outspoken minority. We need to convince them that the conservation role of the zoo is not mere rhetoric.

It is also necessary, however, to realize that the need to conserve and breed animals, the need to maintain them in good health, and the need to exhibit them in naturalistic conditions are mutually reinforcing and place demands upon the knowledge and resources of traditional zoos. Esthetically pleasing and ecologically valid exhibits that are also conducive to breeding and good maintenance are rare. Most zoos have some exhibits they point to with pride. All zoos have ugly, outdated, artificial, unhealthful quarters, odious to animal and keeper alike, that are too often brushed aside. Sometimes, however, what looks nice, what appears natural, may in fact be harmful to the animals. We know of cases where animals bred in atrocious quarters while, at other zoos with "better" facilities, success was rare.

The behavioral researcher can give zoos important cues as to what are the minimum and optimum sized enclosures for maintaining natural groups of animals without the pathological effects of crowding. Temperature, humidity, substrate, structures, and diet can all affect the quality of a display. Habitat groupings, a technique which is very useful since it increases the density of animals per unit area without overly increasing the psychological need for spacing among conspecifics, is an important tool. Yet, such groupings need to be based upon adequate knowledge of natural conditions and careful study in captivity.

WHY BEHAVIORAL RESEARCHERS NEED THE ZOO

In the last few years behavioral research among zoology, biology, anthropology, and psychology departments has become an established and accepted area. This new emphasis has gone beyond the concerns of traditional comparative and learning psychology that, while originally very productive and innovative, relied on too few species studied under artificial conditions, looked at behaviors only marginally important to the animal being studied, and accumulated evidence that indicated that little information is really appropriate to humans. This approach has been pushed into the background, even though it had some important contributions to make (Lockard, 1971; Lester, 1973). Methods developed from this perspective (especially operant conditioning technology) are often valuable and can activate the behavioral repertoire of vegetating humans in institutions and vegetating animals in zoos (Markowitz, this volume).

A MODEL OF THE RESEARCH ENVIRONMENT IN ZOOS

At this point I would like briefly to present a model of the zoo as an institution and research arena (Figure 1). The animal, rather than portrayed in a physical cage, is under the influence of four interdependent factors: the enclosure itself, the maintenance procedures (feeding, cleaning, climate control), the keepers (who carry out the maintenance procedures at the enclosure), and the visitors (for whom the zoo is traditionally operated). The first three factors are characteristic of any animal in captivity, be it on a farm or in a research laboratory. But the unique factor for zoos, aquariums, and wildlife parks are the visitors. Their presence influences and modifies the other three factors and also any research

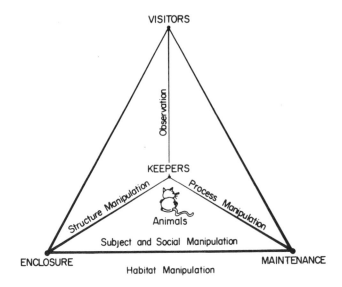

Figure 1
Types of zoo research and the factors influencing them.

endeavors themselves. The fifth interrelated factor, of course, is the animal itself.

The researcher, whether employed by the zoo or not, must consider the pyramid as given. While any behavioral research may affect all five factors, the various classes of research may be broken down into those that primarily operate upon one or two of the five factors, although the whole system is necessarily affected to an extent. Since the purpose of this paper is not to give a detailed overview of zoo research, a brief outline of these kinds of study is in order.

Observation

Observational studies are those involving the description and interpretation of behavior occurring within the zoo environment. This is research that is conducted during short-term studies or for gathering baseline data either for further research efforts or before the zoo alters current exhibits, introduces new arrivals, and so forth. However, even when surreptitious methods are employed, the influence of visitors and keeper is felt. It is frequently necessary to work around peak visiting hours or to restrict visitation to the exhibit to some degree. At the very least visitor behavior must be more carefully supervised—controlling feeding, rock and garbage tossing, glass and bar banging, shouting, and so on. Keepers also need to be informed and appraised of the methods and goals of the study. Indeed, they can be important as resource persons. Modifications in their duties can often greatly aid the study; on the other hand, keepers who feel slighted, imposed upon, or just territorial can vitiate even observational studies. Their cooperation is vital.

Structure Manipulation

Often, exhibit modifications must be made to make a study more feasible. For instance, some enclosures may be so constructed that the entire living area is not visible from the outside. Such barriers as sheds or signs can restrict observation. In a complex social situation where animals may run out of view for a few minutes, such blocked viewing is most frustrating.

Apparatus can also be installed in the cage to facilitate tests of social behavior, learning, discrimination, and the like as in many of the procedures described by Markowitz (this volume). The structure manipulation basically involves modification of the enclosure and the forbearance of keepers.

Process Manipulation

There are studies in which some alterations in the maintenance procedures are made at the most basic level. Included here are changes in the type of food given the animals and adjustment of feeding and cleaning schedules. It may be useful to see how the animal treats natural prey or food. The researcher may also want to have different foods available, not because he is interested in feeding behavior per se, but because such a diet may influence other behavior. For instance, if the animal can get all its nutrition from very concentrated and easily gathered and eaten food, he may have more time available than is characteristic for the species and this might influence social interaction and competitiveness, as Gilbert Boese has reported for the Guinea baboon troop at Brookfield Zoo.

The time of feeding and the amount of food given are also parameters that the researcher may want to manipulate for obvious reasons. The researcher may deprive animals of food for a length of time to increase certain types of social interactions or to facilitate learning experiences. Questions involving deprivation that under some circumstances might be construed as mistreatment need to be evaluated carefully by trained personnel.

The above manipulations are usually relatively long-term in scope but blend over to the more experimental type of situation where objects may be introduced and removed in the animal's enclosure, such as the pioneering studies on curiosity by Glickman and Sroges (1966) at Lincoln Park Zoo or the tool manipulation studies on baboons by Beck at Brookfield

(1973). Another kind of study is the learning or training situation where the animal is expected to alter its behavior through reward or punishment for certain activities. This type of research should be carefully evaluated before it is undertaken.

The researcher interested in learning studies usually has one of two questions in mind. He may be interested in using a training technique to determine a sensory or perceptual ability of the animal; that is, he may want to determine whether the animal has color vision, olfactory discrimination, and so forth. The second type of study involving training is one that focuses on learning itself, such as how fast or how much the animal can learn. Such questions cannot be adequately answered in most zoo situations using the traditional models, methods, and theories of psychology. However, learning studies can be profitably undertaken in zoos, particularly ones that involve rapid 1- or 2-trial learning or the longer term automated efforts described by Markowitz (this volume).

Habitat Manipulation

Habitat manipulation includes all attempts at long-term changes in the environment presented to the animals—changes in climate (temperature, humidity), day–night cycles, vegetation, and physical surroundings. While such changes are related to structure manipulation, they are long-term and mainly influence the enclosure and maintenance factors.

Changes in the physical environment can provide the animal with more things to do and enrich the diversity of behavior, including successful breeding. How do animals actually treat sand, soil, trees, wood and metal perches, plants, etc.? A more complex environment might change or alter the patterns of social formation. We do know that in some lizards, for instance, territory size is a function of structural complexity. Therefore, a more complicated environment enables more animals to be maintained per unit area even though the added physical structures reduce the total amount of room available.

Subject and Social Manipulation

Attempts to alter the inhabitants of the enclosure are important research strategies. Research on physiology, parasites, disease, nutrition, and the internal control of behavior fall into this category. Clearly, much of this type of work is unsuited for zoos. Being alert to unusual occurrences in zoos ("nature's experiments") can be valuable and practical, however, as shown by the baboon visual problems discovered by Beck (this volume) and a similar finding on elephants by Markowitz (this volume).

Often, animals born in zoos cannot be reared by the natural parent. When this happens, as it frequently does, and the babies are handreared, opportunity exists for the gathering of much basic data and for producing a tamed animal that might be very suitable for experimental work and comparison with mother-reared animals. Yet few of these opportunities are taken, outside of brief reports on formulas, temperatures, and so on.

Another useful research strategy is to vary the social composition of exhibits. After a researcher determines a presumed social hierarchy, for instance, it may be illuminating to remove certain animals or introduce new animals and see what effect this has upon the social system. Perhaps breeding behavior will be facilitated or inhibited by keeping pairs together or separated. Similarly, rearing of the young and the interactions of the mother–infant pair with other animals may also be important factors. Particularly timely are studies involving the mutual relations of various species kept together. Ideally this should involve forms living naturally in the same ecosystem (e.g., African savannah, Central American rainforest), but often closely related allopatric forms are kept together, particularly birds and reptiles.

RESEARCH IN SMALL ZOOS

Larger zoos, in general, have responded more quickly and more successfully to new demands with breeding, conservation, and education programs. Smaller zoos also need to show an awareness and action toward these ends. Research can help solve problems that are specific to a given institution, and, while I respect the literature, there are many gaps that need to be filled. It is becoming increasingly apparent that smaller zoos will in the future need to specialize and develop the expertise to breed their stock.

Research problems in the small private or municipal zoo, with a minimal budget, are of a different order than those of the wealthier, larger zoos. First of all, a small zoo has no institutional framework such as ideally exists in a zoo that has specific departments devoted to research or education. At best there may be a curator or director of research who can handle and coordinate problems that arise. If none is employed by the zoo, then the next best alternative is to enlist the aid of some experienced individual at a nearby college or university who has shown some interest in using the zoo as a research and teaching resource. In my own case, a job needed to be done and I stepped forward and offered to do it, frankly with the hope of creating an atmosphere where a full-time position could be created for an individual who would like to base his career largely around research in zoos. But trustees and city council members do not appreciate precipitous action unless it deals with revenue or industry. Besides, one needs to educate the lay leaders to the fact that research on zoo animals is not the same as vivisection.

Publicity concerning successful research efforts should be widespread in the community. Most small zoos are in small cities, the citizens of which like to hear that they were involved in some unique venture. Further, the results of such research should be pointed out in exhibits and signs, as in the now famous pioneering efforts at Brookfield (Rabb, 1969, 1970).

The types of research undertaken at small zoos should focus initially on animals that are relatively common and easier to maintain in larger numbers than are the rare, expensive, and exotic species. It is amazing that so little is known about the behavior of many animals that have been maintained in captivity for hundreds of years. Bears are a good example of this situation.

BEHAVIOR RESEARCH ON CAPTIVE BLACK BEARS

The behavior of many common animals kept in zoos has been sadly neglected. This became apparent to me after my arrival in the Great Smoky Mountains area of Tennessee where I quickly became interested in the native black bear (*Ursus americanus*). Scanty information was available on social behavior, communication, sensory abilities, motor abilities, learning, and behavioral development of bears. The most complete review was Meyer-Holzapfel (1957). When two female cubs about 10 weeks of age were orphaned early in 1970 in the Great Smoky Mountains National Park and became available for study, we dropped most other projects to embark on the work discussed below.

Initially, the bears were handreared in my home (Figure 2). Handrearing is an excellent means of producing tame, amenable subjects as well as producing developmental data. Since the Knoxville Zoo, at that time, was not a very savory institution, it was soon necessary for us, with the mutual cooperation of the National Park Service and the University of Tennessee, to construct an enclosure and related support facilities at a somewhat remote area within the national park (Figure 3). While the relative isolation from disturbance was helpful, the bears were exposed to frequent visitors from the nearby Environmental Education Center (including classes of elementary school children) and a moderately well-traveled hardtop road was on the opposite side of a stream about 100 meters away.

Our studies included all the areas mentioned above, and here I can only present short synopses of some of the studies undertaken by students (e.g., Bacon, 1973; Pruitt, 1974) and myself. The emphasis is purposely on crude, basic, cheap, and seemingly unsophisticated methods for gathering information. Such methods are clearly applicable in many zoos where research resources are limited.

Figure 2
The "play pen" built for housing young bears reared in a human habitation.

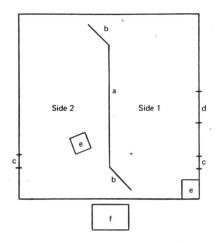

Figure 3
The bear enclosure, 18.3 × 18.3 meters, constructed for the research described here: (a) dividing fence, (b) large gates, (c) personnel gates, (d) drive-in gate, (e) den sites, (f) observation and storage shed.

Weighing and Measuring

Good consistently recorded physical data is important, particularly during major developmental periods. Obviously large, potentially dangerous species present problems. While bears are normally put in that category, Figures 4 and 5 illustrate the weighing methods we were able to use with our bears. Monthly measurements with tape were taken of the distance between the ears, the snout-to-ear distance, front and rear paw lengths, waist circumference, ear-to-tail distance, and height. Since it was noted that the tongue is an important "appendage" in bears, a means of measuring how far a bear could extend it through a fence in order to obtain a favored morsel, such as a raisin, was developed (Figure 6).

Figure 4
Bathroom scale method of weighing bears (up to 40 kg).

Analysis of Basic Behaviors and Activity Patterns

The behavior of bears can be broken down into fairly molar functional units such as walking, running, climbing, fighting, scratching, feeding, sitting, sleeping, and so on. It is possible to monitor the behavior for sessions of 30–60 minutes at various times of day. The use of checklists, tape recorded notes, and videotape monitoring can be very useful (see Hutt and Hutt, 1970). Changes in activities over 24 hours, seasons,

Figure 5
Swing method of weighing bears. Raisins or apple pieces were used to lure bear. Scale was about a meter above bear's head.

Figure 6
Method of measuring tongue protrusibility.

weather, age of bears, and maintenance and structure modifications can also be studied.

One of our observations with implications for maintaining black bears in the zoo is that the young cubs spend a great deal of time in the trees (as compared to adult bears) and prefer to sleep there, even in rain and snow, rather than in a straw filled shelter. Later, they do use the shelter (Figure 7) and then begin digging extensive but shallow depressions in the ground (Figure 8). We covered these with sections of huge logs (Figure 9) or wooden frames (Figure 10). Initially we noted the bears gathering twigs, hemlock branches, and other materials both in open sleeping areas at the base of a large tree and into the dens themselves. We then put straw in a central location, and it was quickly moved into appropriate areas (Figure 10). Figure 9 illustrates a common movement. I have seen brown bears in zoos on concrete spend hours in autumn raking back and forth a few leaves that blew into the cage from the outside.

Film and videotape records allowed one to go beyond the description gathered as it occurs, especially fast paced social interactions and the details of limb and head movements entailed in locomotion, feeding, and

Figure 7
Both cubs in den. Kate manipulating straw.

Figure 8
Excavation of tree roots in one corner
by Kate.

Figure 9
Part of log used to cover hole in
Figure 8 and used as den by Kate.
Note also the nest gathering move-
ments of straw.

Figure 10
Den developed by Kit. Large amount of straw was moved into den from area to left of picture.

digging. It also allowed the accurate gathering of topographical descriptions of the actual sequence and form of motor patterns and communication strategies; here we found super-8 motion picture film to be quite helpful. Generally, the fast black and white emulsions, such as Kodak Tri-X and Kodak 4-X, proved more suitable for our shaded area and dark subjects. However, commercial processing is less available than for color film and there are no financial savings. With normal or slow motion films, it was possible to develop frame-by-frame analysis that became quite complicated and necessitated the use of computers (Pruitt, 1974).

Curiosity and Exploration
The bears (named Kit and Kate) were always interested in many things but were also shy and easily startled. This is not a paradoxical statement. In general, when in relatively novel situations—such as our back yard—dogs, trucks, and other stimuli would quickly send them up trees. However, they generally habituated quickly, more so to trucks than

dogs. Approach of a person at the pens would usually send them to the base of a tree, which they might partially climb until the person was recognized, and then quickly approached him or her.

Such objects as purses, bags, and scarves were readily approached and manipulated extensively. This led to our testing Kit and Kate on a standardized zoo animal curiosity test involving blocks, chains, dowels, and hose developed by Glickman and Sroges (1966). Our data indicated that the bears performed above the average scores for either carnivores (exclusive of bears) or primates, with the chain manipulated most.

Play

The bears engaged in self-play with a variety of objects, conspecific play, and play with their human foster parents (Burghardt and Burghardt, 1972). Conspecific play has been extensively analyzed from filmed sequences and 341 encounters described in the daily logs by Cheryl Pruitt (1974).

The play-soliciting or play-invitation movements included two very distinctive types but was variable and included at least five head movements, ten kinds of locomotor movements, and five paw movements. The roll-over type of play, probably derived from submission, was seen more frequently as a response to humans than to other bears and is illustrated elsewhere (Burghardt and Burghardt, 1972). A variant involved one bear lying on its back and pawing at the other bear as it came near (Figure 11). The jaw bite, directed near the base of the front limb, was seen in both young (Figure 12) and large bears (Figure 13) —note the ear positions and muzzle twist. This would often develop into "jaw wrestling" (Figure 14) and "head jockeying" (Figure 15). Play might be terminated by running away or, more subtly, by turning away (Figure 16). Detailed records of the initiation and termination of play, the direction of bites, and so on were made. A general model of play interaction is depicted in Figure 17.

Agonistic Encounters

It has been claimed by a number of writers on animal behavior (e.g., Lorenz, 1954; Morris, 1965) as well as by people who have actually worked with bears (Krott and Krott, 1962), that ursids are dangerous and unpredictable; this is attributed to their lack of communicational channels, facial musculature, and distinctive coloration. As we raised our bears and interacted with them, it became quickly apparent that we had little problem gauging the mood of the animal or in interpreting some specific movements. The discussion of play certainly indicates this.

Figure 11
Play solicitation by bear in foreground. All ears are in upright position.

Figure 12
Play-bite invitation in mother-reared young cubs.

Figure 13
Play-bite invitation by Kit. Note the muzzle twist by the recipient.

Figure 14
Jaw wrestling form of play.

Figure 15
Head jockeying form of play.

Figure 16
Head turning form of play termination.

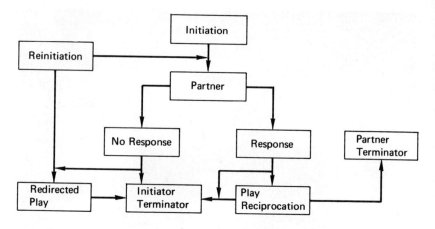

Figure 17
Model of conspecific play interactions.

But the concern about predictability really is based on predicting "aggressive" behavior.

Kit and Kate, as well as other captive bears, engaged in little conspecific aggression. As young cubs, fighting broke out only when one cub drained her bottle first. Even up to the end of the period they were kept together (41 months) they would sit close, head to head, while eating. Only rarely were spontaneous fights seen, and this usually occurred when a single small food item was present. Threat behavior, involving "jealousy" over who received attention from a human (Figure 18), did occur. Our cubs were two females, but a male–female pair we also observed responded similarly, except during the mating season.

In order to gather film records of fights it was necessary to resort to manipulation. One side of the two compartment hog feeder used for dry dog food was lashed down (Figure 19). Since there was room for only one animal to insert his snout, agonistic actions sometimes resulted. From these filmed encounters a series of six highly predictable stages of a fight could be extracted. These involved variety of ear positions, muzzle "expressions," body postures, bites, and vocalizations. It is important to note that vocalizations are a definite sign of aggression since they are absent from all stages of play behavior.

Figures 19–22 illustrate some of these stages of agonism. We have seen similar sequences in fights in other contexts and by different animals, and they are compatible with brief reports from field studies (Jonkel and Cowan, 1971).

Figure 18
Young cub near foster mother signaling sibling to keep her distance.

Figure 19
As bear No. 1 eats at the feeder, No. 2 approaches eliciting the threat "low moan" vocalization in No. 1. No. 2 may threaten by flattening ears and extending the upper lip.

Figure 20
No. 2 directs a bite and/or slap to the rear quarters of No. 1 who continues low moan.

Figure 21
No. 1 whirls around to No. 2 and a short charge involving standing on the rear legs occurs. All this accompanied by loud growls.

Figure 22
Both animals assume a face-to-face head down posture on all fours. Ears continue
flat and the eyes rotate into mutual contact. One animal will eventually break this
confrontation posture and resume eating. The loser leaves the area, perhaps after
emitting a few final low moans.

Feeding and Foraging
The manipulation and ingestion of a wide variety of natural and unnatural food items were filmed and analyzed. Single-frame analysis
allowed precise delineation of the movements involved (Figure 23).

Foraging behavior itself (locating food scattered in the environment)
was also studied. An omnivore does much searching, and this was an
attempt to mimic, at least crudely, the natural situation. A series of
ingestive related behaviors were developed from observation and then
used in a comparison of foraging and "random" exploration as shown
in Table 1.

Food Preferences
More experimental techniques were utilized in an attempt to assess the
food preferences of black bears and whether any stable seasonal or
developmental differences occurred. A modified primate testing apparatus was constructed and hinged to one of the doors on the opposite
side of the gate hinge (Figure 24); hence, the regular gate could be
swung out of the way and the test gate substituted. A paired comparison

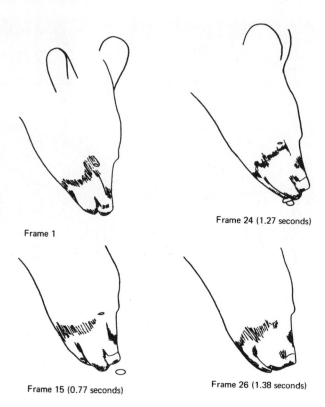

Frame 1

Frame 24 (1.27 seconds)

Frame 15 (0.77 seconds)

Frame 26 (1.38 seconds)

Figure 23
Four frames showing approach and ingestion of an acorn re-drawn from super-8 film. Note lip extension at frame 15.

TABLE 1   Comparison of the Frequency of Foraging Behaviors Prior to Ingestion with Those Occurring Within a Random Sample

Foraging Behaviors	Prior to Ingestion	Random Sample
Smell closely objects in cage	77	11
Smell air	1	9
Smell closely the ground	33	
Apparently random walking	72	15
Direct movement	99	32
Apparent random walking and smell closely the ground	62	65
Smell closely and use front paws	105	6
Consumption of food or water	59	76
TOTAL	508 (71%)	214 (30%)

Figure 24
Food preference test apparatus. After items are prepared out of the bear's sight, the door is raised and the tray pushed through to the waiting subject.

technique was used, each bear being tested on every combination of five native food items and eight nonnative food items every 2 weeks for a year. The nonnative items were added because our bears had experience with them and also because bears in the park frequently seek out garbage and campers' food supplies.

Figure 25 gives the bear's view of the testing. The hinged partition was raised and the selection board pushed forward for every trial. To make a choice, the bear had to push aside the hinged screen lid and remove the food from the clean stainless steel bowl. Side cues were controlled. Figure 26 illustrates the selection board results on native foods for Kate. A sophisticated scaling test (Torgerson, 1958) showed that a good rank ordering occurred. Little effect of season was noted.

Color Vision

Bears are reputed to have poor eyesight. After many false starts, we were able to develop a satisfactory technique for testing visual discrimi-

Figure 25
Bear's view of the food preference test.

nation in bears. Basically it includes a 2–5-item simultaneous choice situation with food reward. The basic apparatus is shown in Figure 27. The plywood item constitutes the olfactory control board (OCB). Many visual experiments with animals neglect to adequately control olfactory cues, and in an animal reported to be highly olfactory we were determined not to make that mistake. All the test stimuli were placed on the OCB's. Under the copper screening all OCB's contained 16 raisins, a favored delicacy. Only the "correct" stimulus had a raisin or two on top of the screen under the stimulus that could be eaten by the bear. Although physically able to demolish the OCB for all the raisins, our bears merely found the correct one and, often checking the others, proceeded to the gate leading to the opposite side. The bears were given a raisin for going through the gate and then keeping on that side until the next test was prepared. In this way 10–20 tests a day could be run.

In order to test for color vision, the bear was first trained to discriminate a gray from a colored cup. Then various shades of gray (13) and various shades of the color (5–9) were used. This is a variation of the classic methods used by von Frisch to show color vision in honeybees. Spectrographic records of all pigments used to color the cups were made. The bears were able to learn the task quickly and did demonstrate discrimination between the colors used. A sample of the data is shown

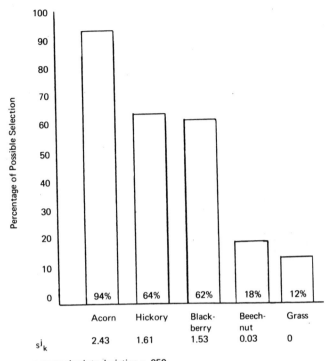

Figure 26
Results of year-long paired comparison ranking of five native foods.

Figure 27
The olfactory control board and stimulus cup used in color discrimination training.

TABLE 2 Results of the Blue-Green, Two-Choice Discrimination for Kate

Date of Test	Stimulus Items		Illumination[a]		Results	Criterion Reached
	Blue (pos.)	Green	Left	Right		
14 May 1972	1,3,4–6	1–5	Shade	Shade	25 of 27	yes
16 May 1972	1,3,4–6	1–5	Shade	Shade	25 of 25	yes
17 May 1972	1,3,5,7	2–5	Sun	Sun	16 of 18	yes
26 May 1972	1,2,5,7	1–4	Sun	Sun	16 of 16	yes
1 June 1972	1,3,5,6	2–5	Sun	Shade	17 of 18	yes
1 June 1972	2,4,5,6	2–5	Shade	Sun	16 of 16	yes
2 June 1972	1,2,5,7	1–4	Sun	Sun	17 of 18	yes

[a] Illumination of the stimuli is relative to the path of approach of the subject.

TABLE 3   Results of Form Discrimination Testing

Type/ Date of Test	Task	Score	Significant Positive Discrimination $p \leq 0.05$
Discrimination			
12 May 1972	▲ vs ●	23 correct of 24 trials	yes
Size Constancy			
15 May 1972	▲ vs ●	10 correct of 10 trials	yes
	▲ vs ●	9 correct of 10 trials	yes
	▲ vs ⬤	9 correct of 10 trials	yes
	▲ vs ⬤	10 correct of 10 trials	yes
Rotation			
16 May 1972	▲ vs ●	10 correct of 10 trials	yes
	▼ vs ●	10 correct of 10 trials	yes
Background Reversal			
17 May 1972	▲ vs ●	10 correct of 10 trials	yes
	△ vs ○	14 correct of 20 trials	no
18 May 1972	▲ vs ●	10 correct of 10 trials	yes
	△ vs ○	9 correct of 10 trials	yes
Object Recognition and Retention			
7 September 1972	▲ vs ●	9 correct of 9 trials	yes
	▲ vs ╱	9 correct of 10 trials	yes
	▲ vs ▮	9 correct of 10 trials	yes
	▲ vs ⊥	9 correct of 9 trials	yes
	▲ vs ◆	9 correct of 9 trials	yes
	▲ vs ▬	9 correct of 9 trials	yes
	▲ vs ■	9 correct of 9 trials	yes
2 May 1973	▲ vs ●	11 correct of 12 trials	yes

in Table 2. Varying sun and shade were also used to eliminate systematic brightness cues.

Form Discrimination

Using a similar technique but employing 5-sided cubes, painted white, with black equal area forms (triangles, squares, circles, etc.), it was found that easy learning and ready discrimination occurred, including background reversal, and constancy effects (Table 3).

WHAT THE BEAR STUDIES TELL US

The above described studies, although only representative of many others, are just preliminary steps in our complete understanding of this species. Our major aim was to explore the feasibility of performing various kinds of studies, rather than to virtually exhaust the myriad of unanswered questions about any given aspect of bear behavior. However, some conclusions relevant to keeping bears in the zoo can be drawn from this research, the aim of which was basic knowledge.

1. Trees are important, and real or artificial ones will greatly add to an exhibit involving young bears.

2. Nesting material should be provided at appropriate times of year.

3. A section of the enclosure should contain dirt. Perhaps various small food items could be periodically hidden therein.

4. Omnivorous bears handle a wide variety of fruits and vegetables in interesting ways. Berries (on branches), apples, acorns, and even watermelon should be tried.

5. In the forest habitat of black bears there are many objects, which bears manipulate and explore. Providing some different objects that they can play with and manipulate will add to the interest of bear exhibits. Except for permanent objects, these should be removed daily and new objects introduced.

6. Keepers can be instructed in the types of bear aggressive behaviors and their intention movements.

It would appear that any zoo with the desire could foster similar studies involving imagination, empathy, humor, a knowledge of research techniques, and most importantly, the flexibility to utilize the results of such research in the design and maintenance of exhibits.

CONCLUSIONS

Research in zoos is traditionally categorized as basic or applied. It is more useful to view research as arising either out of concerns relating to the maintenance, breeding, and effective display of animals or from questions about the animals derived from a consideration of their biology and behavior irrespective of their captive state. As both the zoo and academic approaches become more naturalistic, mutual benefits should increase and small institutions will find the fostering of research more valuable and feasible.

REFERENCES

Bacon, E. S. 1973. Investigation on perception and behavior of the American black bear, *Ursus americanus*. Doctoral dissertation, University of Tennessee, Knoxville, Tenn.

Beck, B. B. 1973. Cooperative tool use by captive hamadryas baboons. Science, 182:594–597.

Burghardt, G. M., and L. S. Burghardt. 1972. Notes on the behavioral development of two female black bear cubs: The first eight months. Pages 207–220 in S. Herrero, ed. Bears—Their biology and management. IUCN, new series No. 23. International Union for Conservation of Nature and Natural Resources, Morges, Switzerland.

Glickman, S. E., and R. W. Sroges. 1966. Curiosity in zoo animals. Behaviour 26: 151–188.

Hediger, H. 1950. Wild animals in captivity. Butterworths, London.

Hutt, S. J., and C. Hutt. 1970. Direct observation and measurement of behavior. Charles C Thomas, Springfield, Ill.

Jonkel, C. J., and I. McT. Cowan. 1971. The black bear in the spruce-fir forest. Wildl. Monogr. 27:3–57.

Krott, P., and G. Krott. 1962. Zum Verhalten des Braunbären (Ursus arctos L. 1958). Z. Tierpsychol. 20:160–206.

Lester, D. 1973. Comparative psychology. Alfred Knopf, New York.

Lockard, R. B. 1971. Reflections on the fall of comparative psychology: Is there a message for us all? Am. Psychol. 26:168–179.

Lorenz, K. 1954. Man meets dog. Methuen, London.

Meyer-Holzapfel, M. 1957. Das Verhalten der Bären (Ursidae). Handbuch Zool. 8(Part 10, 17):1–28.

Morris, D. 1965. The mammals. Harper & Row, New York.

Pruitt, C. H. 1974. Ontogeny and communication in the black bear Ursus americanus. Doctoral dissertation, University of Tennessee, Knoxville, Tenn.

Rabb, G. B. 1969. Educating the zoo public about animal behavior. Pages 14–17 in The use of zoos and aquariums in teaching animal behavior. American Association of Zoological Parks and Aquariums, Wheeling, West Virginia.

Rabb, G. B. 1970. The unicorn experiment. Curator 12(4):257–262.

Torgerson, W. S. 1958. Theory and methods of scaling. John Wiley & Sons, New York.

Zuckerman, S. 1932. The social life of monkeys and apes. Routledge, London.

III

REPRODUCTIVE BIOLOGY

HERMAN M. SLATIS

Department of Zoology, Michigan State University
East Lansing

The Genetics of Inbreeding

Any zoo that tries to develop a breeding program will soon be faced with the problem of inbreeding. Although the basic stock will usually consist of an unrelated pair of animals, daughters will generally have no other prospective mates than their own father or a brother and inbreeding will seem to be a necessity. Our culture prohibits most forms of close inbreeding in man. Our attitudes toward inbreeding in animals are determined to a great extent by our attitudes toward human inbreeding. Because we are wary of inbreeding, there is a tendency to drop a program containing inbreeding after any minor setback. In contrast, many modern societies looked upon inbreeding with favor, and at times the ancient Egyptians looked upon brother–sister marriage as particularly desirable.

Although inbreeding may be dangerous, it can also be useful in a number of ways. It is a tool that should be encouraged where useful.

DEFINITIONS

There are several possible definitions of inbreeding, but it is herein defined as any breeding between close relatives. Inbreeding increases the probability that a particular gene received from one parent will be matched by an identical gene from the other parent. An offspring that has received a pair of identical genes is *homozygous* for the gene pair. Outbred individuals will often receive different genes from each parent and will have the minimum homozygosity for the species. The amount by which inbreeding increases the probability that a gene pair will be homozygous can be predicted from a statistical analysis of the pedigree. This increase in homozy-

137

gosity is described as the *inbreeding coefficient, F* (Wright, 1969) (see Figure 1). If F is low, or about 0.03, there is little inbreeding. F is only 0.0625 for a first cousin mating, but it is 0.25 for a brother–sister or parent–offspring mating. In other words, the offspring of a brother–sister mating will have about 25 percent more of their gene pairs homozygous than will an average outbred individual. Except for laboratory rodents, F values rarely exceed 0.50 in mammals.

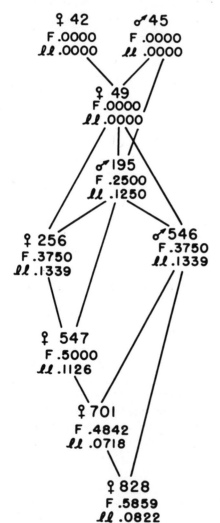

Figure 1
Pedigree of female 828, a highly inbred European bison. The inbreeding coefficient, *F*, and an alternative inbreeding coefficient, *ll*, are shown for each animal.

ADVANTAGES OF INBREEDING

Inbreeding is usually practiced due to one or several of the following advantages.

Convenience

Inbreeding is particularly convenient for the rare species found only in widely scattered zoos. Where the alternative is either the transportation of a prized and perhaps bulky specimen or the development of a method of artificial insemination, inbreeding solves far more problems than it creates.

Maintaining a Special Strain

Inbreeding may be useful for producing and maintaining animals of a desirable strain. For example, if one were to capture an albino in a species in which they are rare, the only simple method for getting and maintaining a stock of albinos would be through inbreeding for several generations. Most other color variants would follow the same rule. Very rare geographic varieties may also have to be maintained in zoos by inbreeding. Often, a specimen of a rare variety will be outbred and then its descendants will be selected based on those that conform to the idealized type for the variety. The result may agree in one or two characteristics such as size or color, but the remaining genes, which may have contributed to the ecological adaptation of the variety, will often be lost.

Detecting Undesirable Genes

Inbreeding provides a method for detecting the presence of rare genes that produce severe defects when homozygous. In cattle, if a bull is expected to be used for extensive breeding, he is often tested by being bred to a number of his own daughters (Wriedt and Mohr, 1928). If no abnormal offspring appear, he is then outbred. In the past, many widely used bulls have later been shown to have spread a gene that produced stillborn or deformed calves over many generations. Eventually, the bull's descendants have to be screened for the gene or have to be discarded as potential breeders. A program of inbreeding may be particularly important for zoos because of their tendency to rely on a single male to breed with several females.

Producing a Uniform Strain

Uniform strains are best developed through inbreeding. While this is desirable for many economic and experimental purposes, it probably is of no importance for zoos.

DISADVANTAGES OF INBREEDING

The undesirable effects of inbreeding may occur at any time in the life cycle, beginning with fertilization and ending with death in old age.

Fertilization

Taillessness in mice can be caused by a complex gene that enables a sperm carrying that gene to be more effective in fertilizing an egg than is the sperm not carrying it (Dunn, 1964). Genes affecting sperm activity may be widespread, but they are generally not detectable. The tailless "gene" is readily observed only in the presence of a gene that has not been seen in natural populations. When homozygous, the tailless gene causes death or male sterility so that this particular strain of mice is not homozygous for long. Genes of this type could block an inbreeding program effectively, and male sterility does seem to be a common finding in inbreeding studies.

Spontaneous Abortion

Many different genes have the ability to cause death between fertilization and birth. There is evidence from both cattle and man that inbreeding causes a high loss of embryos soon after conception (Mares et al., 1958; Schull, 1958; Slatis, 1963). In man, inbreeding has only a small effect on survival late in pregnancy, on stillbirth, or on death immediately after birth (Slatis, Reid, and Hoene, 1958). The data for cattle do not distinguish between these periods, but the effect of inbreeding apparently is much greater than occurs in man. Zoo pedigree books do not distinguish between an abortion, stillbirth, or death on the day of live birth, and do not include information about obvious malformations associated with these deaths.

Juvenile Death

Death during the juvenile period is the most readily detectable effect of inbreeding in man and other mammals. This type of death, in which the youngster often wastes slowly away, is sociologically and psychologically the most traumatic effect of inbreeding. This trauma may bias the observer disproportionately against the further use of inbreeding. In a zoo, when a particular pair of animals has produced such an affected offspring, the obvious remedy is to find a new mate for each of the parents. However, since the offspring of zoo animals are themselves zoo animals, this remedy merely postpones the problem for a later generation. The undesirable genes remain in the zoo. It would be advantageous to use a system of inbreeding and testing to develop a stock from which the undesirable gene was eliminated. The stock would be stronger for eliminating the gene.

Vigor
Most studies of adult vigor have been of economically important traits such as milk production in cattle. Because these traits have long histories of artificial selection, it is unreasonable to extrapolate from them to other characteristics.

Fertility
In the European bison, Slatis (1960) found that adult cows, whether inbred or outbred, had about 0.463 probability of having a calf in a given year. This delicate measure showed no difference due to inbreeding. Bowman and Falconer (1960) reported that inbreeding led to sterility and the loss of nine of ten lines of mice. However, the surviving line had the high fertility of the outbred ancestors. A line of beagles that I have worked with quickly achieved inbreeding values of 0.75 without other signs of decreased vigor, but some of the highly inbred males have been infertile.

Death
There is a good deal of evidence for the role of inbreeding in this process, and it may be relatively minor.

INBREEDING AND OUTBREEDING

The European bison were divided into two subspecies based on location, one in the central European lowlands and a rarely observed subspecies limited to the Caucasus Mountains. Only one Caucasian animal was ever bred in the lowlands, and his descendants proved to be highly successful. One of his two daughters produced 20 single-born calves in a period of 21 years, a record for the species. This is an eloquent testimony to outbreeding for one of the most outbred females ever born in this species. In the next few generations, the inbred descendants of this same Caucasian bull had a very high frequency of stillbirth and early death, and they were disproportionately responsible for the bad effect of inbreeding on early death rates in the species (Slatis, 1960).

A plan that emphasized some inbreeding would identify those individuals that have undesirable genes and remove them from the breeding program leaving the stock sturdy. This appears to be what happened to the European bison during the 1930's. The weakest stocks were reduced in numbers and some lines died out; the strongest stocks proliferated and gave the species the survival margin that it needed. Zoos will be tempted to compare their inbred generations with the outbred generations that preceded them, but putting off inbreeding in species where new stock is not readily available will only postpone and perhaps magnify some problems.

In general, we have very little information about the extent of inbreeding in natural populations. In most species, males must wander from their birthplaces before maturity, but it is possible that females tend to stay near their mothers throughout their lives. In species with harems, young females may frequently mate with their own father. There may also be a tendency for young half-siblings from the same harem to mate. If inbreeding is frequent in a species, genes that cause serious defects will tend to be lost quickly. Under conditions of general outbreeding with only accidental inbreeding, such genes would tend to accumulate. The average number of undesirable genes per animal would be much higher in outbreeding species than in inbreeding species. The outbreeding species would have a high probability of juvenile death when inbred in zoos, while inbreeding species would have far less juvenile death.

In addition to individual genes that lead to abnormality and death, many geneticists believe that homozygosity is bad in itself. They argue that inbreeding will, by increasing homozygosity, cause severe effects. Analysis of the data for the European bison, however, suggests that their continued inbreeding might be successful because the bad effects of inbreeding were more closely related to the action of single genes that to F, the measure of how homozygous the animals were (Slatis, 1960). Similarly, in domesticated species it has been shown that selection can control some of the bad effects of inbreeding.

The various pedigree books for wild-caught animals and their captive-born descendants would serve their purposes better if they included data on all matings, cases of apparent impotence and infertility, early abortions, causes of death, and similar demographic features. However, even with nothing more than the parentage and the dates of birth and death, the pedigree books provide data that cannot be obtained in any other way for an analysis of the nature of genetic variability in mammals. I, for one, am greatly indebted to the zoo breeders for their efforts in keeping and publishing accurate records.

SUMMARY

Inbreeding has four advantages: convenience, producing and maintaining a desirable strain, detecting undesirable genes, and producing a uniform strain. The most obvious disadvantages are the loss of juvenile animals and the possible reduction of fertility in highly inbred animals.

Analysis of the vigor of endangered species would be facilitated by putting more detail into the pedigree books.

REFERENCES

Bowman, J. C., and D. S. Falconer. 1960. Inbreeding depression and heterosis of litter size in mice. Genet. Res. 1:262–274.

Dunn, L. C. 1964. Abnormalities associated with a chromosome region in the mouse. Science 144:260–263.

Mares, S. E., A. C. Menge, W. J. Tyler, and L. E. Casida. 1958. Some sources of variation in conception rate and pregnancy loss in parous Holstein cows. J. Anim. Sci. 17:1217.

Schull, W. J. 1958. Empirical risks in consanguineous marriages: Sex ratio, malformation, and viability. Am. J. Human Genet. 10:294–343.

Slatis, H. M. 1960. An analysis of inbreeding in the European bison. Genetics 45:275–287.

Slatis, H. M. 1963. Problems in the study of consanguinity. Pages 236–243 in E. Goldschmidt, ed. The genetics of migrant and isolate populations. Williams and Wilkins, Baltimore.

Slatis, H. M., R. H. Reis, and R. E. Hoene. 1958. Consanguineous marriages in the Chicago region. Am. J. Human Genet. 10:446–464.

Wriedt, C., and O. L. Mohr. 1928. Amputated, a recessive lethal gene in cattle: With a discussion on the bearing of lethal factors on the principles of livestock breeding. J. Genet. 20:187–215.

Wright, S. 1969. Evolution and the genetics of populations, Vol. 2. University of Chicago Press, Chicago.

KURT BENIRSCHKE, M.D.
Department of Reproductive Medicine
University of California, San Diego
La Jolla

Cell and Sperm Banks for the Zoological Parks

With some species facing extinction despite numerous and heroic efforts to protect them, the opportunity to gain some knowledge of their physiology and genetics dwindles. It is clearly most desirable that no additional animal species should disappear and, ideally, small groups should be saved for ultimate study. Should such efforts fail it would be useful to possess remnants of such species, other than their skeletons, with which to answer questions of scientific interest. Cell cultures, frozen blood, and sperm are currently suitable examples; perhaps blastocysts and other live tissues can be saved in the future. The point is that it is important to commence now with the systematic efforts to collect and freeze under controlled conditions such materials as become available from appropriate species.

Maintenance of cell cultures, sperm, blood, and so forth in the frozen state is feasible. When cryoprotective agents are employed and controlled freezing is performed, such living cells can be stored in liquid nitrogen, presumably indefinitely. It has been possible to unfreeze, under rigidly controlled conditions, tissue cultured cells and spermatozoa 10 and more years after freezing and observe new viability. Cells will grow and replicate in culture, sperm will fertilize. The cryoprotective agents vary, glycerol and DMSO are most widely used for cultures, glycerol and glycerol–egg–yolk concentrate for sperm, among others. Techniques for freezing are fairly standardized and, once in liquid nitrogen, little if any loss of viability takes place. The greatest loss of cells is incurred during the freezing process and, of course, all safety measures must be employed

144

to prevent accidental unfreezing through loss of liquid nitrogen (cracking of the vacuum lined containers) during storage.

A further aspect of importance is the proper identification (genetic characterization) of tissues frozen for long term. Thus, if such banks will ever serve their ultimate purpose, the records kept must identify, with a great deal of certainty, the nature, origin, age, and other details relating to the animal from which the tissues have been collected.

TISSUE CULTURES

In the practice of mammalian tissue culture, cell lines are established which, under good conditions, will have up to 50 generations. A large number of cells whose nuclei contain the faithful copy of the genome of the animal in question thus becomes available. Most laboratories engaged in such studies make it their practice to freeze away a portion of such cultures from early divisions. This ensures against the loss of lines by laboratory accident (contamination, loss of power, poor nutritives), and recourse to these vials of frozen lines is a common practice in such laboratories. Moreover, after study of some enzymes, chromosomes, or growth characteristics, for instance, it becomes apparent that different types of experiments are indicated. Stock can then be unfrozen; it can be shared with investigators in other laboratories; and, more importantly, it can be saved for the development of new techniques at some future date.

An example is appropriate from recent experiences in cytogenetics. It is easiest to propagate mammalian cells from fibrous tissue, small biopsies collected at death (often hours or days after death), or during surgical manipulations. During cell divisions the chromosomes become condensed and can then be studied in detail. In the past it has been possible only to enumerate chromosomes, make measurements, and outline fairly simple descriptions. From such studies an enormous amount of insight has nevertheless been gained into taxonomic relationships of mammals, leading to the annual publication of a mammalian chromosome atlas (Hsu and Benirschke, 1967–1974). Stock of the valuable animals collected in this effort (approximately 400 species) was frozen. In recent years entirely new techniques became available, which revolutionized our understanding of the organization of the chromosomes (Nilsson, 1973). Special stains allow the recognition of bands and other structures in all chromosomes that are highly specific for a species or, at times, even for individual animals. With these techniques it has become possible to distinguish the chromosomes of some subspecies whose phenotypes are often difficult to identify. More importantly, in attempting to reconstruct evolutionary relationships between species, the location, size, and distance of these bands allows one to make precise judgment as to the nature of chromosomal rearrangements that took place coincident with speciation. In this reconstruction the availability of some frozen stock of animals that rarely become available is of great help in the new comparison of chromosomes.

A specific other example may be cited. In a newborn chimpanzee, McClure *et al.* (1969) identified an autosomal trisomy; the animal had 49 instead of 48 chromosomes and the phenotype was similar to that of human "mongolism." The latter condition is due to trisomy 21, and very recent studies show that this responsible element can be distinguished from chromosome 22, which otherwise appears identical in regular stains. The trisomic chimpanzee baby has since died, but a culture strain has been frozen. Unfreezing this strain at this time would allow one to employ the new chromosome banding techniques to identify whether the anomalies in chimpanzee and man are identical. No such opportunity would ever arise again and the availability of a frozen stock of this culture is thus of inestimable value.

One cannot rationally anticipate future use of frozen cells since it is impossible to forecast new techniques. One can hypothesize that ultimately, as in amphibia, nuclei transplantation may become feasible in mammals. Perhaps such cells will then find uses in substituting for nuclei in zygotes. At present this is science fiction; but it nevertheless compels a systematic effort for the preservation of cell lines, particularly from vanishing species. Clearly, the best place for collection and banks would be the zoo.

BLOOD BANKING

It is presently possible to freeze fresh blood with similar cryoprotectives. The technique is well advanced in man, and recently it has been suggested that rhesus monkeys be employed for the systematic study of the survival of transfused red blood cells after frozen storage (Button, Garcia, and Kevy, 1973). It can be envisaged that such blood be stored for investigative purposes (e.g., enzymes, blood group markers) and also for therapeutic measures. At present many apes and other newborn animals are cared for in zoo nurseries. Installing newborn intensive care nurseries for rare species is also under discussion. Occasionally, transfusions are required for critically ill or injured animals. At our zoo a young pigmy chimpanzee needed a transfusion but blood was unavailable and the animal was not typed. As a result, our collection of pigmy chimpanzees has recently been blood grouped, and we learned that the animals at Antwerp had been similarly grouped (Moor-Jankowski *et al., 1972*). This effort not only delineates fascinating aspects of blood group relationships among chimpanzees and man but will also, it is hoped, lead to the judicious storing of some chimpanzee blood in the customary pellet form. It is also anticipated that it will identify human blood types that may be used for emergency transfusion.

Blood grouping is far advanced in some domestic species, such as cattle and horses. One can readily envisage that it could become feasible in vanishing species, e.g., the Przewalski horse. Thus, blood grouping could become helpful for studbook purposes, tracing relationships, for evolutionary questions, and so forth. Preservation of quantities of red blood cells may thus be of considerable interest to zoological gardens in the future.

SPERM BANKS

An extensive effort has been made to preserve sperm for artificial insemination in numerous domestic animals and in man. The literature on this subject was reviewed comprehensively in a recent conference (Blandau, In press). Preservation of spermatozoa in liquid nitrogen employing various cryoprotective agents is possible in cattle, rhesus, man, and other species for very long periods of time, perhaps forever. It can be unfrozen and used for artificial insemination and retains many normal biologic activities although some are damaged or even destroyed in the process of cooling. In some species, however, many efforts have not been rewarded with success. Presumably, the right cryoprotective agent has not been discovered and continued research is needed in this area.

In general, sperm is collected from artificial vaginas or by electro-ejaculation. The means of animal sedation or immobilization, frequency of electric stimulation, size and types of electro-ejaculator, and means of immediate treatment (dilution, warmth) of semen vary widely among species. We have attempted collection in lion, cheetah, llama, white-lipped peccary, cat, bobcat and a few other species and found that the type of anesthesia has a profound influence, as does vigor and frequency of electric stimulation. While it is easy to create erection under most circumstances, successful ejaculation is not always achieved. Therefore, an alternate route is being employed, the collection of sperm from epididymes in freshly dead specimens. This has been done in a variety of rare bovids and could readily be practiced in most zoos to achieve sufficient knowledge as to how to handle the sperm of any given animal. Freezing and thawing with recovery of viability (mobility) has thus been achieved with sperm of white-bearded gnu, among other bovids.

Mobility alone, however, is an insufficient criterion for the ultimate efficacy of sperm (Friberg and Gemzell, 1973); it rests on the ability of fertilization. Virtually no progress has been made in this respect with zoo specimens. Yet, the availability of effective frozen sperm collected from freshly dead Indian rhinoceroses, for example, would surely be one step toward the ultimate conservation of this species. When different sexes of animals live in different countries and cannot be shipped and they are so rare as to prohibit transportation, it can be envisaged that attempts at artificial insemination may be a valuable tool for successful reproduction. Indeed a few such attempts have not only been suggested but even tried— so far, unsuccessfully. It is therefore most gratifying to learn that at Toledo Zoo the first chimpanzee baby was delivered after artificial insemination (Hardin, 1973). What is more important, the semen was collected from a 31-year-old nonbreeder. Although it was not frozen semen, it is a first step demonstrating the possibilities of the future.

It is of parenthetic importance to mention that the genotype of any semen donor be carefully documented. For instance, in the experience of Swedish investigations (Gustavsson, 1971), a large percentage of Swedish cattle are carriers of a chromosomal translocation because one of the few bulls used for the collection of semen carried such a translocation. This is clearly undesirable.

Ideally preserved semen, however, does not guarantee success. Much too little is known of the basic reproductive biology of most zoo species that are in danger of extinction to assure successful insemination. It is necessary to understand how to ovulate an animal successfully with reflex ovulation, to time the insemination with presumed or known reproductive cycles, to have enough semen available for repeat insemination, and so forth. In this respect a great deal of knowledge exists with some domestic species and many laboratory rodents, but little systematic inquiry has been made in the very divergent species housed in zoos that are ultimately to benefit from such procedures. Thus, basic studies of all aspects of reproductive physiology on zoo animals must precede any hope of this achievement of artificial insemination. With the exception of the Institute at London, no such efforts are conducted in zoos. This is most regrettable and in need of correction for it is unlikely that these parameters of reproduction will become understood through studies conducted at universities, with the possible exception of a few isolated species. Certainly these parameters will not be worked out for those animals in greatest danger since they can only be studied at zoos. It is recommended that sperm conservation of those endangered males be one of the first steps toward preservation of the particular species.

ZYGOTE PRESERVATION

It has recently become possible to preserve young fertilized mouse eggs (morulae) (Whittingham, Leibo, and Mazur, 1972). They may be unfrozen, transplanted to foster mothers and grow into embryos of a different genotype than that of the foster mother. There are interesting biological problems with this system (Uphoff, 1973); nevertheless, it is an exciting new achievement. Newspaper accounts report successful transplantation of a frozen cattle embryo, and the future will doubtless see the exploitation of this experimentation in the elective breeding of valuable livestock. Similarly, conservation of some rarer species' zygotes can ultimately be envisaged, but more basic understanding of reproductive mechanisms needs to be achieved before it will be a practical occurrence to ship blastocysts across the world rather than live animals.

It is idle at this point to discuss this aspect further. Suffice it to say, research in reproductive physiology is advancing at a fast pace in those species that are of interest to man as a researcher or for his financial and nutritive gains. No *a priori* reason exists why the same should not be true of the zoo captives to whom we owe similar responsibility, providing similar efforts were made.

RECOMMENDATIONS

This very brief review and my long time association with various zoological parks suggest to me that it is of urgent importance that in a number of such institutions research efforts of much greater magnitude be established, particularly in the broad field of reproductive physiology. Through what may be termed "basic" studies—for example, endocrinology, cytogenetics, and pathology—the insight gained with respect to human reproduction has been enormous in the last decade. As a consequence, the practice of obstetrics (i.e., *reproduction,* in zoo terms) has undergone dramatic changes to the benefit of the consumers, mother, and infant. There is every reason to believe that the same would be the case in most species, to the ultimate goal of conservation of all species now extant.

From a practical and scientific point of view, compelling arguments exist to promptly establish centralized cell and semen banks as a first measure. Because of the need for proximity to the species to be studied, they must be at zoos. Their goal should be central collection of valuable cell lines and sperm research into optimal collection, and dissemination of such precious stock to qualified workers in other institutions.

REFERENCES

Blandau, R. J., ed. In press. The biology of aging gametes. S. Karger, Basel.

Button, L. N., F. G. Garcia, and S. V. Kevy. 1973. The rhesus monkey as a model for evaluation of the preservation of stored whole blood. Transfusion 13:119–123.

Friberg, J., and C. Gemzell. 1973. Inseminations of human sperm after freezing in liquid nitrogen vapors with glycerol or glycerol–egg–yolk–citrate as protective media. Am. J. Obstet. Gynecol. 116:330–334.

Gustavsson, I. 1971. Distribution of the 1/29 translocation in the A.I. bull population of Swedish red and white cattle. Hereditas 69:101–106.

Hardin, J. 1973. A world's first. Am. Assoc. Zool. Park Aquariums Newsl. 14(8).

Hsu, T. C., and K. Benirschke. 1967–1974. An atlas of mammalian chromosomes, Vols. I–VII. Springer-Verlag, New York.

McClure, H. M., K. H. Belden, W. A. Pieper, and C. B. Jacobson. 1969. Autosomal trisomy in a chimpanzee: Resemblance to Down's syndrome. Science 165:1010–1012.

Moor-Jankowski, J., A. S. Wiener, W. W. Socha, E. B. Gordon, and J. Mortelmans. 1972. Blood groups of the dwarf chimpanzee (Pan paniscus). J. Med. Primat. 1:90–101.

Nilsson, B. 1973. A bibliography of literature concerning chromosome identification—with special reference to fluorescence and Giemsa staining techniques. Hereditas 73:259–270.

Uphoff, D. E. 1973. Maternal influences on mouse embryos and preservation of mutant strains by freezing. Science 181:287–288.

Whittingham, D. G., S. P. Leibo, and P. Mazur. 1972. Survival of mouse embryos frozen to −196° and −269°C. Science 178:411–414.

R. M. F. S. SADLEIR

Department of Biological Sciences, Simon Fraser University
Burnaby, British Columbia, Canada

Role of the Environment in the Reproduction of Mammals in Zoos

Animals living in captivity are maintained in extremely different environments than those occupied by the same species in the wild state. The timing of breeding of such species, and their physiology and behavior, have resulted from the selective pressures exerted from environmental factors over many millions of years. In attempting to provide a suitable environment for the breeding of captive mammals, zoo biologists have recognized that it is impracticable in almost all cases to try and duplicate the entire natural environment in the captive situation. Instead, there is a need to recognize the essential components of this environment and then implement these factors into the captive area to ensure breeding.

The concept of necessary and sufficient criteria for establishing causal relationships between environmental factors and breeding success should perhaps be considered at this point. One can consider a certain environmental factor as being *necessary* for breeding if, when it is absent from the zoo situation, breeding never occurs. A somewhat ridiculous example would be the presence at one time or another of both sexes. More subtly, the need of certain essential vitamins in the diet can be considered as a necessary factor in breeding most mammal species. The zoo biologist, faced with a poor breeding situation, must therefore recognize which of these necessary factors are absent from his particular situation and attempt to correct for them. Environmental components can be *sufficient*

I wish to thank Claus C. Mueller for kindly commenting on the manuscript in an early draft.

151

to ensure breeding if breeding always occurs when certain factors (or more usually a collection of factors) are operating. Perhaps the main problem in many zoo breedings is the difficulty of recognizing, when breeding has occurred, exactly what these sufficient factors were. In other words, a difficult-to-breed species may breed in captivity, but the exact reasons as to why breeding was successful remain unrealized. As a result such breeding successes may become rare or even unique events. The early history of breeding of the cheetah (*Acinonyx jubatus*) (Rawlins, 1972) is an example. Many reports of successful breedings in various zoo journals unfortunately show no analysis of the crucial environmental factors that resulted in birth and rearing of young.

Basically, successful and repeated breeding is recognized as the main criteria of good zoo husbandry and, to many people, it is considered a vital justification for the existence of zoos. This paper will review the effects of some of the main environmental factors of breeding in captive exotic mammals. I have considered elsewhere (Sadleir, 1969a) the role of such factors in wild and domestic mammals. I will here attempt to categorize environmental parameters in such a way that a zoo biologist, faced with the absence of breeding in a species, may be able to consider his individual situation and recognize from these categories which of the necessary and/or sufficient environmental factors are missing.

PHOTOPERIOD

In the majority of temperate and polar mammal species there is good evidence that a seasonally altering daylength regime is necessary to stimulate breeding activity at specific times of the year. The movement of such species to localities nearer the equator usually places them under natural light regimes that are still stimulatory, so that the absence of the correct photoperiod is unlikely to be a cause of cessation of breeding in these cases. There is very little good information on the role of photoperiod in breeding of tropical and equatorial species. It should be remembered that the absence of a definitive breeding season in its native latitude does not necessarily mean that a species may not respond to photoperiod in more northern climates. The recent paper by Strahan, Newman, and Mitchell (1973) comparing the breeding season of 30 mammal species in Sydney, Australia, and London, England, shows this point quite clearly. Manipulation of photoperiod enables the zoo biologist to extend the period of the year during which mating could occur.

Photoperiod only affects the first stage of the reproduction process in all except a few groups of mammals. It is known in some mustelids that photoperiod regulates the end of the delayed implantation period of pregnancy. This may also occur in certain seals. It is possible that a failure to recognize this phenomenon may have contributed to the relatively poor breeding records of some of the exotic mustelids.

Despite the above comments it is unlikely that photoperiod, except in a few special cases, could be considered a repeated cause of poor mammal breeding in most zoo conditions.

NUTRITION

An adequate nutritional intake has been long recognized as a necessary environmental factor in the breeding of zoo mammals. At different stages of the reproductive process female mammals have different metabolic requirements. All competent zoo keepers have recognized for years that the pregnant female needs an increased calorific and protein intake, but the degree of increased intake in late lactation does not seem to be generally appreciated. Work on domestic species and on wild rodents (Sadleir, 1969b) has shown that gross food intakes of up to four times that of the nonreproductive adult are required during lactation. The advent of multivitamin and multimineral diet additives has removed the necessity to determine which of the vitamins or minerals were necessary and sufficient for breeding in most species of mammals.

The role of changes in diet in stimulating reproduction in mammals has been discussed by zoo biologists for years. The Radcliffe–Wacker-nagel–Hediger controversy (Hediger, 1969) with regard to the psychological effects of a varied diet is well known to zoo biologists. However, there is some evidence in the field literature that changes in nutrition may stimulate estrus. This is certainly true for lagomorphs, macropods, and certain tropical rodents. Sheepmen have known for years that "flushing," a sudden increase in the quality of food, will increase the twinning rate, but there appears to be no evidence that flushing can stimulate estrus. In fact, this is unlikely in a species such as the sheep, which is highly photoperiodically responsive. In an evolutionary sense one would not expect a surge of nutritional intake to stimulate breeding in large mammals, because the resulting pregnancy and lactation could result in a subsequent birth at a nutritionally unfavorable time of the year.

Many mammals live in habitats where there is a regular seasonal alteration in the protein:calorie ratio of the dietary intake with or without seasonal variation in total intake due to changes in availability. Under such conditions there are seasons of fat deposition and seasons of fat depletion. It seems reasonable to suggest that a sufficient body fat content may be a necessary prior requirement for pregnancy and lactation.

Nothing seems to be known as to the role of fat balance in relation to gametogenesis, although in many species late pregnancy and particularly late lactation are periods of severe fat depletion. In zoos the diet does not generally vary in a seasonal manner, so that many mammals maintain a constant body fat proportion. It would be of great interest to attempt to seasonally alter the fat balance by manipulating the protein:calorie ratio of the diet to determine if this could improve breeding.

SOCIAL ENVIRONMENT

It is probably fair to say that the main breakthroughs in breeding of mammals in zoos over the past decade have resulted from a better understanding and consequent manipulation of the role of conspecifics in the breeding process. This has resulted from the development of ethology as a discipline and as a result of the recent remarkable increase in behavioral field studies of many African species. It has become possible for zoo biologists to attempt to duplicate the breeding social unit found in the wild. Actually, in many cases, such as with the larger primates, ethological observations in zoos have greatly complemented field studies. Works such as those of Eisenberg (1966) and his many subsequent publications on social organization and Fraser (1968) on reproductive behavior in ungulates have proved invaluable to zoo biologists in understanding the environment necessary for breeding many social mammals.

Despite these ethological advances, a major difficulty has arisen in recognizing the necessary and sufficient components existing in the social environment that result in breeding. All zoo biologists are aware of the many reports of unexpected birth or unpredicted copulation. Even in groups such as the larger Felidae, which have been maintained in zoos for hundreds of years, there is still remarkable lack of agreement in zoo reports as to the effect of the presence of the male on the successful mothering of cubs.

To manipulate the social environment in zoo species costs much more than the provision of a total diet which, although it may contain superfluous items, does contain all the necessary components. Few zoos can afford to set up various combinations of conspecifics to determine arbitrarily the best social arrangement for breeding. There would seem to be a need for ethologists to develop ethographic models of the major zoo species that could describe the necessary and sufficient social components for breeding in captivity. If this were possible, even on a probabilistic basis, it could provide an enormously valuable tool to all zoo biologists interested in improving their breeding records. The exciting inclusion in the latest edtion of the *International Zoo Yearbook* of records of multiple generation births makes the construction of such ethograms now possible for a number of mammal species in captivity. One could therefore hope that in time the *large* zoo literature on hand-rearing techniques would eventually diminish.

PHYSICAL ENVIRONMENT

The dimensions and nature of the cage or pen surrounding mammals in zoos are of a unique importance to their breeding. There have been a number of dramatic successes in breeding once some necessary component of the physical environment was recognized and compensated for. The provision of isolated heated dens for pregnant polar bears (*Thalarctos maritimus*) and the need to provide a nonslippery padded flooring during parturition and early suckling of giraffe (*Giraffa camelopardalis*) are two good examples. Many European zoos have successfully bred the larger deer after inhibiting the potential damage of the rutting buck. This is done by dividing the enclosure with a fence with narrow openings (*hochzeitsgatter*), which slows down the following male.

A final example of the physical dimensions of the environment affecting reproduction in zoos is described by Walther (1961). During copulation the males of certain ungulates, particularly *Oryx* species, assume a vertical orientation with their front legs well above the female's back. This necessitates high ceilings in their cages, as the long horns of such species otherwise impede mating. Often the necessary physical environment overlaps with the required social environment in that territorial space or sites for scent marking are required for complete mating ceremonies.

Two final points remain for general consideration. Firstly, if "spontaneous" or "unsolicited" breeding successes are to be utilized as indicating the correct environment for breeding in a species, every attempt should be made to recognize the necessary and sufficient components operating at that time and to report them subsequently in the literature. Many zoos do not have staff with the necessary biological background to make such recognition possible. Thus, it becomes very necessary when breeding successes are reported to give as full and exact a description as possible of the particular cage or enclosure and of the husbandry regime employed during conception, gestation, and lactation. As such reports accumulate in the literature, zoo biologists will be able to recognize with increasing certainty the necessary and sufficient environmental components for successful breeding.

REFERENCES

Eisenberg, J. F. 1966. The social organizations of mammals. Handbuch der Zoologie 8:92.

Fraser, A. F. 1968. Reproductive behaviour in ungulates. Academic Press, New York and London. 202 pp.

Hediger, H. 1969. Man and animal in the zoo: Zoo biology. Seymour Lawrence/Delacorte Press, New York. 303 pp.

Rawlins, C. G. C. 1972. Cheetahs (*Acinonyx jubatus*) in captivity. Int. Zool. Yearbk. 12:119–120.

Sadleir, R. M. F. S. 1969a. The ecology of reproduction in wild and domestic mammals. Methuens, London. 321 pp.

Sadleir, R. M. F. S. 1969b. The role of nutrition in the reproduction of wild mammals. J. Reprod. Fert. Suppl. 6:39–48.

Strahan, R., P. J. Newman, and R. T. Mitchell. 1973. Times of birth of thirty mammal species, bred in the zoos of London and Sydney. Int. Zool. Yearbk. 13:384–386.

Walther, F. 1961. Mating behaviour of certain horned animals. Int. Zool. Yearbk. 3:70–77.

DEVRA G. KLEIMAN

Reproduction Zoologist, National Zoological Park
Smithsonian Institution, Washington, D.C.

Management of Breeding Programs in Zoos

Over the past twenty years, public discussion concerning the development of effective breeding programs in zoos has become increasingly more sophisticated. No staff member of a modern zoo would dare to say publicly that the best method for breeding a particular species is to "put a male and female together and let nature take its course." We know that "nature" does not always produce viable young, while some prior planning and a little knowledge might.

To date, most efforts in the development of breeding programs have been applied to the rare and endangered ungulates, e.g., Père David's deer, *Elaphurus davidianus*. Several of these programs have been successful. However, it is likely that project proposals that prove that a zoo can produce and maintain viable breeding populations of particular endangered species will eventually be required before a zoo will be granted permission to obtain specimens.

Even though most staff members in a zoo know what type of information is necessary in order to initiate an effective breeding program, zoos seem to have difficulty both in the collection of the relevant information and in its later application. Zoo personnel tend to treasure knowledge gained from their own experience while bypassing information arising from the experience of others, even though there are organizations, such as AAZPA, and periodicals, such as the *International Zoo Yearbook,* whose

I am grateful to H. Buechner for his comments on the manuscript. I would especially like to thank the keepers and volunteers from the National Zoo who have helped to show that the regular collection of behavior data in zoos can contribute significantly to the development of breeding programs.

express purpose is to disseminate information. Zoos still make management decisions and build exhibits with surprising disregard for the experience and knowledge of other zoos and scientific researchers. To some extent, this situation exists because not all zoos see the breeding of their specimens as a major commitment. However, even in institutions where breeding has a high priority, the same tendency exists.

In addition to this problem there are still large gaps in our knowledge of reproduction and of the environmental requirements for propagation of the majority of species being maintained in zoos (Sadleir, this volume). Unfortunately the availability of certain modern techniques, such as artificial insemination, has lulled some zoo personnel into a false sense of security, believing that the technical skills of endocrinologists, physiologists, and embryologists can solve problems of infertility as if by magic. By adopting the fashionable attitude that viable captive populations can be achieved by mixing sperm and eggs and placing the mixture in the womb of a closely related species (Francoeur, 1972), the zoo staff are avoiding their own responsibilities. Advances in the physiology of reproduction undoubtedly will prove to be a tremendous benefit to zoos; however, at present our knowledge of reproduction in most species is too meager to allow zoos more than the occasional use of some of the new techniques (Benirschke, this volume). For example, it is impossible to successfully artificially inseminate a female when the characteristics of the estrous cycle are not known for that species and methods of ovulation induction have not been perfected.

For the majority of zoos, true advancement in the development of effective breeding programs will be achieved only with increased understanding of basic reproductive processes in the numerous species maintained and with a continuous exchange of information and experience among zoo staffs.

Two of the best examples of this type of cooperative information exchange are found in the Reproduction in Captive Cheetah Seminar held at the October 1972 AAZPA meeting and the conference on saving the lion marmoset (Bridgwater, 1972). The results of these meetings represent a pooling of the knowledge of experts on reproduction, behavior, and management of cheetahs and marmosets, and they provide a baseline for any zoo desiring to breed these species as well as an example for future meetings concerned with the captive propagation of individual species. Physiological techniques were not viewed as a sole solution to infertility; the emphasis in both meetings was on developing proper management procedures.

In developing a successful breeding program, there are problems at many different stages of the reproductive process as well as in managing the results of breeding. I am going to review each stage separately since each presents its own special obstacles; methods for determining reproductive condition will be discussed as well as factors that might naturally or artificially influence it. I will assume that nutrition and the basic housing of species are adequate for reproduction. The discussion will deal exclusively with mammals

THE ESTROUS CYCLE

One of the greatest barriers to breeding involves insufficient knowledge of the reproductive cycle of the female, both in terms of the length and type of estrous cycle and in the determination of heat. In mammals, there are three basic types of cycle.

In several orders, females experience a period of heat only once each year, but stay in heat for several days, e.g., most canids, some insectivorous bats, some mustelids, and pinnipeds (Asdell, 1964). Such species are said to be *monoestrus* and are usually seasonal breeders. If mating and fertilization do not occur at the appropriate time, reproduction is delayed a full year.

Many mammals show seasonality in breeding, but have several estrous cycles or full reproductive cycles (e.g., mating, pregnancy, birth, lactation) during a season. Females that are seasonally polyestrus are found among macaque monkeys (*Macaca* sp.), many rodents, ungulates, and some mustelids. In other groups, especially in the relatively constant conditions of captivity, estrus may recur periodically throughout the year, e.g., anthropoid apes, ungulates, caviid rodents, and felids (Asdell, 1964).

The degree to which a species exhibits a fixed breeding season varies. Many captive and wild mammals may breed year round, but the majority of births occur during a limited period of several months. It is generally recognized that captive animals, like domesticated species, may breed more frequently, mature earlier, and have larger litters than their wild counterparts (Hediger, 1965).

For many mammals, we either know or can partly predict the basic type of reproductive cycle. However, the specifics of the cycle are often unknown, e.g., the cheetah (*Acinonyx jubatus*) probably exhibits heat recurrently as do most other large felids, but the length of the cycle has not been established nor have methods been developed for successfully predicting the occurrence of heat. Wolves (*Canis lupus*) are monoestrus and breed in late winter, but predicting the exact date of heat in a particular female is impossible.

Some methods for the determination of the occurrence of heat and the length and type of the estrous cycle are discussed below.

In some mammals, there are easily visible morphological changes that precede or accompany the onset of heat. Wolves, like dogs, exhibit a bloody discharge from the vulva approximately 1–2 weeks before they are receptive; therefore a careful examination of the anogenital area of wolf females beginning in early January should signal the close onset of heat. Females that do not exhibit a discharge may not be reproductively active. Similarly, most caviomorph rodents have a vaginal closure membrane which becomes perforate before heat (Weir, In press); catching females every day will provide information on the cycle.

In several species of primates (Hafez, 1971), the genital area of the female swells conspicuously during the week preceding ovulation, and menstruation occurs about 2 weeks later. Maintaining records of the cyclic occurrence of swellings and menstruation (and their disappearance to determine whether a female is pregnant) will keep staff apprised of the reproductive condition of females. Similarly, in many mustelids, the vulva swells prior to breeding (Wright, 1963). Simple charts can be developed for recording these changes and keepers can fill them in on a daily basis. If personnel incorporate this into the regular routine, then a daily check requires less than a minute per female.

There are other methods of determining heat that can be carried out by the keeper and curatorial staff, although with greater time output. In all mammals, behavior changes accompany the onset of heat. Although some changes are very subtle, systematic observations of specimens improve considerably the chance of discerning estrus, especially where males and females may be maintained separately except for purposes of breeding.

For some specimens, the presence or absence of a single behavior pattern may indicate receptivity, e.g., one tigress (*Panthera tigris*) at the National Zoological Park regularly displayed the mating posture (lordosis) to humans when in heat. However, for most mammals, several behavior patterns increase or decrease in frequency as estrus approaches. Moreover, changes in the behavior of males housed with or close to a female may signal changes in her reproductive condition. By closely following the behavior of individuals, it is possible to anticipate when a pair should breed or when a male and female should be introduced for breeding.

In order to use behavior to determine reproductive condition and therefore to successfully diagnose causes of infertility, the method of data collection must be standardized and observations must be performed on a regular basis. At the National Zoological Park, we presently use checksheets for observations on individual specimens. Both keepers and volunteers (from Friends of the National Zoo), mostly with no formal training in behavior, participate. Check-sheets have been used for watching Indian rhinoceroses (*Rhinoceros unicornis*), several species of marmoset (*Saguinus oedipus, Callithrix geoffroyi*, and *Leontopithecus rosalia*), tigers, cheetahs, lesser pandas (*Ailurus fulgens*), and giant pandas (*Ailuropoda melanoleuca*). Although the reasons for observing particular animals vary, we have found the check-sheets to be a useful aid in the determination of heat [Kleiman, In press(a)]. Basically, they provide series of observations that can be used as a basis for comparison with the behavior exhibited by the same animal under different conditions or by different specimens of the same species.

A check-sheet is developed from reviewing the literature on the behavior of a particular species as well as combing the experience of keepers, curators, and scientists. A set of 10–15 behavior patterns which might vary with the physiological or social condition of the animal is chosen. For example, in felids, the occurrence of rubbing, rolling, urine-spraying, and calling are obvious choices. In social species maintained in a group, anogenital sniffing by males, mutual grooming between pairs of animals, frequency and kind of aggressive behaviors, and scent marking are all behaviors that are likely to vary with reproductive state. The males of some species may show "flehmen," circling and driving the female. The check-sheet can be developed so that behavior in isolation may be recorded as well as behavior with a potential mate. The list of behaviors that we have been using for female cheetahs and the final female cheetah check-sheet is presented in Table 1.

In filling out a check-sheet, the observer should record how often and for how long the particular behavior patterns occurred. During a set observation period, for example, a cheetah might roll over three times, urinate once, and rub her cheek on the wall four times. Since these are behaviors of short duration, only the number of occurrences need be recorded. For behaviors that occur less frequently but last longer, a duration measure is useful. For example, an increase in general activity, as exhibited by pacing, may signal the onset of heat. Our Indian rhinoceros female paced almost continuously for a full night a few days before her first successful copulation. Thus keeping records of the length of pacing bouts is useful. Moreover, social interactions with other cagemates, such as mutual grooming in monkeys, should be recorded in terms of how long the individuals continued the activity.

In order for the recorded data to be comparable from one day to the next, it is essential that a standard set of rules be developed for the observations.

1. The length of the watch should be fixed; 30–60 min/d should be sufficient for many species. Ungulates, however, require longer periods of observation, e.g., 90–120 minutes.

2. The watch should be conducted at the same time every day so that observed variations in behavior are real and not due to changes in the

TABLE 1 Check-Sheet Used for Recording Behavior Data on Cheetah (*Acinonyx jubatus*)

FEMALE CHEETAH BEHAVIOR		
Date.................................... Time.................................... Weather....................................		
Behavior	Bouts	No. of Times or Duration
Rubbing: cheek		
forehead		
shoulder		
flank		
Rolling		
Anogenital Grooming		
Calling: purr		
yipe		
stutter		
growl		
Urinating: squat		
spray		
Pacing		

GENERAL COMMENTS: (running, stalking, etc.)

timing of the watch. A specimen will often behave quite differently before as compared with after feeding. Observations should be conducted when the animals are most active.

3. As few people as possible should be involved in recording the behavior of a particular specimen since apparent behavior changes may be caused by differences in observers rather than the animals. Moreover, individual specimens may habituate to the presence of one or two people more easily than a constantly changing set of observers.

If a master score sheet is maintained of the check-sheet records, changes in patterns should begin to emerge. A decision to introduce a pair of animals for breeding can be made more securely when records are carefully maintained and a standard form used. Another benefit to using a check-sheet and insisting that behavior observations be considered part of the daily routine is that observers become more sensitized to subtle changes in the specimen's behavior.

In analyzing the results of the observations, it should be remembered that the behavior of an animal in heat will change in a relative rather than absolute sense. Moreover, each specimen will have certain individual peculiarities, but in summarizing the results even a gross inspection of the data may provide relevant information. In Table 2, the occurrence of five behavior patterns in a tigress, Mohini Rewa, are presented for a 4-month period. Only the presence or absence of a pattern during a 6-day period is noted. Using more detailed methods of analyzing the data [Kleiman, In press (c)], we decided that Mohini had experienced four major heat periods, in early December, mid-January, early February, and early March. The same trend can be seen in Table 2.

There are several more complex methods for determining heat cycles in females, which require the cooperation of specialists as well as increased manipulation of the animal. For example, using the vaginal smear technique, cyclical changes in the structure of the vaginal wall, produced by hormone changes, are analyzed by studying the numbers and types of cells that are sloughed from the vaginal surface. A smear may be obtained by gently scraping the wall of the vagina or introducing a pipette into the vagina and withdrawing some vaginal fluid. In the latter method, saline may be used if the vagina is very dry. The smear is then put on a microscope slide and stained; cell types can be viewed under a microscope.

Vaginal smears should be taken on a daily basis and thus require regular handling of the animal. They are successfully used to determine estrus in many laboratory rodents and dogs, but the results are less consistent in farm animals, e.g., cows, mares, and ewes (Cupps, Anderson, and Cole, 1969). With respect to exotic mammals, vaginal smears of Asiatic elephant, *Elephas maximus* (Jainudeen, Eisenberg, and Tilaker-atne, 1971), Indian rhinoceros, noctule bats, *Nyctalus noctula* (Kleiman and Racey, 1969), and many species of New and Old World monkeys (Evans and Goy, 1968; Rosenblum, 1968; Hafez, 1971; White, Blaine, and Blakley, 1973) have been investigated, but rarely have smears been utilized as a sole method for determination of the estrous cycle. In some cases they were of little or no value in predicting heat. In Old World monkeys and chimpanzees (*Pan troglodytes*) much more emphasis has been directed to correlating mating and ovulation with changes in the visual appearance of the sex swelling (Hafez, 1971; Graham et al., 1972). Of course, even if a daily smear is not useful for predicting heat, it may be valuable for indicating the occurrence of copulation since sperm will be found in the vaginal fluid after mating.

Another more sophisticated technique for analyzing the estrous cycle is based on the finding that human basal body temperature rises on the

TABLE 2   Occurrence of Four Different Behavior Patterns in the Tigress, Mohini, During Estrous Cycle

Date	Roll	Flank Rub	Cheek Rub	Call	Urine Spray
1972					
Dec. 1	+		+	+	
Dec. 7	+	+	+	+	
Dec. 13	+			+	+
Dec. 19			+	+	+
Dec. 25	+			+	+
Dec. 31	+			+	+
1973					
Jan. 6	+	+		+	
Jan. 12	+		+	+	
Jan. 18	+	+		+	+
Jan. 24				+	+
Jan. 30					+
Feb. 5					+
Feb. 11	+		+	+	+
Feb. 17					+
Feb. 23	+			+	+
Mar. 1	+	+	+	+	+
Mar. 7	+	+			
Mar. 13				+	+
Mar. 19	+				+
Mar. 25			+	+	+
Mar. 31				+	+

NOTE: Cross indicates that a behavior was observed at least once during a 6-day period.

day of ovulation. This technique is used as a means of birth control (Hafez, 1971), but has not been generally applied to other mammals. In a recent study of the pig-tailed macaque (*Macaca nemestrina*), daily temperature measurements, using telemetry, did not reveal a consistent rise in temperature at ovulation nor through the latter half of the menstrual cycle (White, Blaine, and Blakley, 1973). From determinations made with a rectal thermometer, 5 of 12 squirrel monkey (*Saimiri sciureus*) females in a relatively small cage showed an 8-day cycle (Gould, Cline, and Williams, 1973). In both studies, there was considerable individual variability in the temperature change. Whether this method can be widely applied to other mammals is open to question.

Finally, estrous cycles can be studied by analyzing changes in hormone levels either in the blood or in urine. With the recent development of several techniques for directly analyzing small quantities of hormones, hormone assays have become common in numerous research laboratories. Laboratory rodents have been studied extensively, but investigators have also been examining hormone levels in domestic farm animals (Hansel and Echternkamp, 1972), ferrets (Blatchley and Donovan, 1972) and Old and New World primates (Hopper and Tullner, 1969; Graham *et al.*, 1972; Stabenfeldt and Hendrickx, 1973; Preslock, Hampton, and Hampton, 1973). The use of such techniques for determining estrous cycles in exotic mammals is possible, but will probably not become commonplace in the near future. As with temperature changes, there is variability in individual hormone levels, and trends can only be seen by pooling the results from several animals. Moreover, the routine collection of blood or urine is a serious obstacle to the use of this method.

Most morphological, physiological, and behavioral methods for the determination of estrus are complicated by the fact that there is considerable individual variability. Thus, the use of more than one method would certainly increase chances for estrus detection. Even more important, flexibility must be maintained when deciding which methods should be applied to a particular problem species or specimen. The National Zoo's Indian rhinoceros female was tame enough to allow vaginal smears to be taken regularly by keepers. The same technique would have been impossible with our black rhinoceros female. Similarly, collecting urine from a rhinoceros may be easier than from a pipistrelle bat (*Pipistrellus* sp.).

NATURAL CONTROL OF FEMALE REPRODUCTION

Numerous factors have been shown to affect female reproductive cycles (Hediger, 1965; Sadleir, 1969). They include climatic variables, e.g., light, rainfall, temperature, and humidity, amount of space, types of social stimulation, nutrition, and many species-specific factors. A disturbance of a single variable may upset reproduction and, for the majority of mammals, we are ignorant of the relative importance of each variable in reproduction (Sadleir, this volume).

At present, there is sufficient knowledge to provide many species with a climatic environment grossly similar to that found in the natural habitat. Modern zoos are now building exhibits in which light cycles, temperature, and humidity are carefully regulated. Unfortunately, financial limitations prevent zoos from maintaining all specimens under such controlled conditions. Nor is it necessary; many animals can adapt and breed in captive situations that only vaguely approximate the norm for the species. However, it is essential to be constantly aware of the environmental needs of particular species, if breeding is the desired goal.

The amount and type of social stimulation required for successful reproduction has been discussed before (Hediger, 1965; Sadleir, 1969), but will be briefly reviewed here since this is one factor affecting reproduction that is not regularly considered. Regardless of whether a species is seasonally monoestrus or polyestrus, the presence or absence of conspecific(s) of the same or opposite sex may be essential in stimulating reproduction.

For example, in mice (*Mus musculus*) and deermice (*Peromyscus maniculatus*), it has been shown that the presence of an adult male (or even just his odor) will initiate and synchronize the estrous cycles of groups of females who have been isolated from males; in female mice exposed to males or male odor, puberty is advanced. Recent studies of such widely separated species as cuis (*Galea musteloides*) and swine have shown similar effects (Eisenberg and Kleiman, 1972).

The mere presence of a male in the vicinity, however, may be insufficient to detect heat or to promote successful reproduction. Frequent tactile contact may be necessary to stimulate the female (and male) to the point of breeding. Our Indian rhinoceroses had unsuccessfully attempted to copulate several times under conditions where they had limited access to one another. Successful copulation finally occurred after they were given 1½ months of constant contact with frequent interaction. Previous problems in breeding many large mammals may be due to insufficient tactile contact between a pair.

The presence of other females in heat may also provide stimulation, especially in species which are polyestrus. Schaller (1972) and Kleiman [In press (c)] have noted that lion (*Panthera leo*) and tiger females, in olfactory contact, often show signs of heat during or shortly after another female's heat.

For colonial species, stimulation from both sexes may be essential in promoting breeding. As has been mentioned by Hediger (1965) and Perry, Bridgwater, and Horseman (1972), far too often pairs of animals are maintained where a group is what is needed.

The opposite requirement may, of course, hold true, i.e., crowding or constant contact with conspecifics may inhibit reproductive cycling. This has been shown for heteromyid rodents, which live in normally dispersed populations (Eisenberg and Isaac, 1963).

The long-term association of several specimens, e.g., where younger animals reach sexual maturity in the presence of adults of the same sex, may cause reproductive inhibition, especially in species where a dominance hierarchy is formed within each sex class, e.g., *Canis lupus* (Rabb, Woolpy, and Ginsburg, 1967), marmosets (Epple, 1970); and dwarf mongooses (*Helogale parvula*) (Rasa, 1973). This inhibition may be direct, i.e., physically preventing an animal from copulating (the wolf) or subtle in its appearance (marmosets). For these species, only the alpha female may have young. Maximum reproduction can only be achieved by periodically removing the maturing offspring.

Even where all females in a group cycle and reproduce, eventual infertility may result if (1) the density of specimens reaches a certain saturation limit (Sadleir, 1969) or (2) the group or pair is never split or provided with new blood. Infertility through the latter condition (known as "familiarity breeds contempt") can be prevented by periodic exchanges of specimens among zoos.

The final consideration in providing an appropriate social environment for reproduction concerns the development of pair preferences or bonds, such as are seen in wolves (Rabb, Woolpy, and Ginsburg, 1967) and marmosets (Epple, 1972). Disruption of pairs may temporarily or permanently inhibit reproductive cycling in a female.

ARTIFICIAL CONTROL OF FEMALE REPRODUCTION

Infertile females may be unresponsive to all efforts to induce breeding through natural means, such as altering the environment, improving nutrition, and changing the social milieu. In such cases, it might prove useful to attempt to induce heat and ovulation artificially through the use of hormones. Advancement in this research field in recent years has occurred mainly with farm animals as a means of controlling fertility and synchronizing the breeding of females (Polge, 1972; Hafez, 1968) and with nonhuman primates (van Wagenen, 1968; Ovadia et al., 1971; Dukelow, 1971). Gonadotropic hormones are regularly used to improve the fertility of women (Lunenfeld, Insler, and Snyder, 1972), but typically cause multiple births. There has been only sporadic research on ovulation induction (with natural or artificial matings) in other mammals, e.g., chinchilla (Chinchilla laniger) (Weir, 1973), rabbit (Adams, 1972), cat (Colby, 1970), and even the giant panda (Bramwell, Rowlands, and Hime, 1969). At the National Zoological Park, we have made a few preliminary attempts to induce heat and mating in infertile specimens, e.g., the binturong (Arctictis binturong) [Kleiman, In press].

Although inducing ovulation using hormones is not often attempted in zoos, it should eventually prove possible both to manage the timing of breeding and to solve certain fertility problems using this technique.

MATING BEHAVIOR

When it is known that a female is exhibiting normal heat cycles, and has a mate but does not become pregnant, attention should be paid to the copulatory behavior of the pair. The basic copulatory patterns of mammals have recently been reviewed by Dewsbury (1972). Abnormalities in sexual behavior may be aided by re-pairing the individual with a proven mate.

Mammals are divided into spontaneous ovulators and induced ovulators, depending upon the degree to which copulatory stimulation is essential for ovulation. Known induced ovulators are found among the insectivores, rodents, carnivores, artiodactylans, and lagomorphs (Jöchle, 1973). However, the dividing line between the two mechanisms of ovulation is not always clear, suggesting that there is a gradation in the response to coitus (Conaway, 1971). For this reason, zoo personnel should always allow the maximum amount of copulatory contact between males and females who must be introduced for breeding. It has been shown for rats that insufficient mating stimulation has an adverse effect on the maintenance of pregnancy and litter size (Wilson, Adler, and LeBoeuf, 1965). In species that copulate many times during the female's heat, insufficient copulatory stimulation may prevent pregnancy altogether, as has been suggested for the tiger [Kleiman, In press (c)].

ARTIFICIAL INSEMINATION

Under conditions where a mature male is unavailable for breeding or progeny of a specific lineage are required, artificial insemination may be desirable. As already mentioned, it cannot be used successfully unless ovulation can be induced in the female or the time of natural ovulation can be predicted accurately.

Jones (1971) and Benirschke (this volume) have recently reviewed the present state of artificial insemination technology. More specific reviews have been presented by Dukelow (1971) for nonhuman primates and by Foote (1969) for farm animals. Although the use of this technique has expanded dramatically over the past few decades, artificial insemination has been applied almost exclusively to livestock breeding; thus, methods of semen collection, sperm storage, and insemination have all developed using farm animals.

Certain methods may be transferred for use with zoo animals, e.g., electro-ejaculation would be an easier method for semen collection than training a male to an artificial vagina, especially since zoos are unlikely to need a constant flow of semen for insemination, at least in the near future.

One of the greatest problems limiting the widespread use of artificial insemination by zoos is the difficulty of preserving semen for long periods, while still maintaining the viability of the sperm (Jones, 1971). This is especially true because demand for the technique by zoos arises mainly when a fertile male is unavailable; thus, semen cannot be collected and used immediately, but must be diluted, preserved, and transported. Jones (1971) describes the numerous methods of diluting and freezing semen; clearly, this is still an area of active research, and for most species no single method has been found that gives consistent results.

Because of the many drawbacks, artificial insemination is not a technique that will be widely used by zoos in the near future, except in selected cases. We should expect success in those mammalian groups closely related to species with which research scientists are already working, e.g., nonhuman primates, ungulates, and some carnivores (felids and canids).

DETERMINATION OF PREGNANCY

Several of the methods used to detect estrus may be valuable in determining when pregnancy occurs, e.g., in chimpanzees or macaque monkeys, cyclical changes in the sexual swelling may disappear. At the same time, visual inspection of the animal may show an increase in weight and development of the mammary glands. Changes in the female's behavior may be noticeable if check-sheet data from the postmating period are compared with earlier results. If weighing is practical, weight increases will often indicate that the female is gravid.

A common method of determining pregnancy in farm animals and nonhuman primates is by palpation (Hafez, 1968; Hendrickx and Houston, 1971). The female must be immobilized for a proper examination; this may be accomplished with a hardy animal using a squeeze cage or handling. Palpation may be impractical, however for more delicate species, unless drugs are used for immobilization. Since a positive result may be obtained on a single examination, pregnancy diagnosis by palpation is feasible for many species in which regular handling for taking vaginal smears or blood samples to determine estrus would be inadvisable.

Finally, testing for concentrations of chorionic gonadotropin in urine or plasma and progesterone in plasma using both biological and chemical assays (Hampton, Levy, and Sweet, 1969; Hendrickx and Houston, 1971; Hodgen et al., 1972) could be a useful method of pregnancy diagnosis. Such methods are regularly used at the National Zoo for determining pregnancy in the great apes (Tullner and Gray, 1968).

Knowledge of a female's pregnancy is important for zoos mainly so that adequate time is available to prepare an environment that is conducive to birth and lactation. Zoo personnel frequently find themselves in the embarassing position of announcing a birth one day and a death the following day since pregnancy was not suspected and no preparations had therefore been made. A significant number of newborn young die in zoos because of disturbance to the female by staff trying to correct for this lack of readiness or by other specimens cohabiting with the mother.

The regular screening of females for pregnancy would reduce the incidence of neonatal deaths considerably since personnel could plan for the birth. The importance of detecting pregnancy has been largely ignored in zoo breeding programs.

DELAYING PHENOMENA IN PREGNANCY

For a majority of mammals, fertilization and implantation occur shortly after ovulation and copulation; pregnancy thus proceeds in a steady fashion. However, there are several species, distributed in a seemingly random manner throughout the taxonomic orders, in which fetal development may not follow soon after mating. Numerous species show the phenomenon of delayed implantation in which the implantation of the blastocyst is delayed. Included are such diverse animals as ermine, *Mustela erminea*; roe deer, *Capreolus capreolus*; and red kangaroo, *Megaleia rufa* (Sadleir, 1969). For some species, the delay occurs during lactation, and implantation will occur soon after the nursing litter is weaned or lost. For others, the delay is fixed in with the annual cycle and appears to function so that both mating and birth occur under optimum conditions.

Two other delaying phenomena, having a similar function to seasonally fixed delayed implantation, are (1) delayed fertilization, where sperm are stored in the uterus for several months after mating but before ovulation (temperate–climate insectivorous bats), and (2) delayed development, where implantation occurs but embryonic growth is retarded during the winter months (leaf-nosed bat, *Macrotus californicus*) (Bradshaw, 1962).

In all three phenomena, where the delay is seasonally fixed, such environmental factors as light and temperature exert a direct effect on the course of pregnancy. If a breeding program is planned for such a species, the zoo must have knowledge of and control over these variables. For example, in order to successfully breed temperate climate insectivorous bats, the animals must be maintained for 5–6 months in a state of torpor. One method involves storing animals in small cages in a refrigerator during the winter months with limited food, but plenty of water (Kleiman and Racey, 1969). The bats can be put into hibernation in the autumn and removed the following spring, with periodic checks during the winter to ensure that animals are not exhibiting too rapid a weight loss. The exhibit colony can then be on view to the public during the busy spring and summer months.

Other species that have delaying phenomena but do not hibernate, like ermine, still require natural temperature and light changes for reproduction; in temperate zone zoos, they should be left outdoors all year round.

PARTURITION AND REARING

Although half the battle is won once a female is pregnant, success in breeding only comes when the offspring of a female breed and rear their own young. Zoos often do not achieve this goal because of mistakes made around the time of parturition.

Environmental conditions are extremely important during this sensitive period. The females of many mammals require absolute isolation during birth and the early stages of rearing, e.g., polar bears (*Thalarctos maritimus*) and probably cheetahs. If birth and early nursing are to be monitored, a closed circuit television or a tape recorder must be set up well before the expected date of parturition. Kühme (1973) recently followed the early development of fennec foxes (*Fennecus zerda*) by placing a tape recorder in the den.

Where isolation from humans and conspecifics is only necessary during parturition, a partial shelter may be constructed from which the female can emerge and return to the social group. A natural social milieu should always be sought. For example, lactating females of species in which the male (or social group) participates in rearing should not be isolated. In fennec foxes (Kühme, 1973), marmosets (Bridgwater, 1972), and green acouchis (*Myoprocta pratti*) (Kleiman, 1970), an isolated mother has poorer reproductive success.

Except for species where the young are born in a very precocious state and follow the mother from birth (many ungulates) or where the young are carried by the mother until they are relatively independent (most primates and many marsupials), mammals need a shelter or nest in which to rear the young. The type of shelter, its size, and location in a cage should be based on the known natural history of the species. Nest material should be provided, if necessary. If a female is being left with conspecifics, extra shelters should be provided. For example, lactating tree shrews (*Tupaia belangeri*) leave the young alone in a nest, and continue to interact and sleep with the male. Young are typically destroyed unless two nesting boxes are provided (Martin, 1968). All of these requirements should be available to the female well before the expected birth date so that disturbance is minimized as a female nears term.

The onset of parturition is often difficult to detect. A female may spend more time in the nest or exhibit a restless pacing. On the other hand, there may be no obvious signs of an impending birth. The recording of behavior using a check-sheet may indicate that parturition is imminent more reliably than casual irregular observations. When the date of parturition can be predicted, a "no disturbance" policy should be instituted and persons unknown to the female kept out of her vicinity, especially with species or specimens which are known to be sensitive. It is rarely necessary to have more than two keepers keeping track of a birth; if trouble arises, specialists can always be called in.

Once labor begins, checks on a female should be kept to a minimum and conducted wherever possible without her knowledge. Although this may appear to be common sense, human anxiety often overrides consideration of the female; well-intentioned people may do more harm than help during a birth.

Parturition should proceed smoothly if the environmental conditions are adequate; when it does not, the zoo veterinarian may have to help deliver or, if a labor is prolonged, do a Caesarian section.

The most crucial period for the survival of young is the 2–3-day period immediately after birth. The occurrence of lactation and the development of a nursing routine can be monitored with a check-sheet using indirect methods of observation; these should indicate whether the process is developing smoothly. Such measures as nursing frequency, time with the young, grooming the young, etc., can be recorded.

Although this early nursing stage is crucial for the development of the mother–young bond, we have little knowledge or control over this phase of reproduction. Due to our own inadequacies, there is a tendency to be pessimistic where the competence of the mother is in question. As a result, American zoos hand-raise hundreds of infant mammals every year. Since the pioneering work of Harlow (1971) and other scientists, it has become common knowledge that the infants of some species when raised in isolation from conspecifics display inadequate social, sexual, and maternal behavior when adult. The more severe the isolation, the more disturbed is the reproductive behavior. That American zoos continue to hand-rear young at the slightest suggestion of difficulty implies that the tradition is strong and not easily changed. A policy of leaving young with the mother (especially with a primiparous female) except in an emergency should be part of the philosophy of any breeding program. If young must be hand-reared, especially for social mammals, a zoo should try to find a compatible adult or young as a companion for the infant.

MALE REPRODUCTION

Compared with the female, little attention has been paid to methods of assessing male reproductive activity. There are two reasons why male reproductive processes have been largely ignored: (1) One good male (or his semen) is capable of fertilizing numerous females and inadequate males can therefore be discarded, and (2) there are fewer stages in the physiology of male reproduction where difficulties can arise.

However, males (or a particular male) may be indispensable in a breeding program, especially if the female's reproductive success depends upon the active participation of the male. As already mentioned, a sexually active male may be necessary in order to induce both heat and ovulation in a female. Moreover, in several species, e.g., marmosets, the male is intimately involved with the rearing of the young. Thus, the possession of a male with either a physiological or behavioral disorder can completely disrupt a breeding program.

There are several methods of assessing male reproductive activity that, when combined, should be highly predictive. Testes size can be visually inspected and compared with other males. In many mammals, males (like females) undergo seasonal changes in gonadal development that correlate with the onset and waning of sexual activity (Sadleir, 1969). These can be charted in daily records.

In some mammals, changes in testes size are accompanied by other morphological changes. The seasonal shedding and development of antlers in deer is one example. Squirrel monkey males display "fatting," whereby increased development of the forearms and shoulders is seen.

The behavior of a male is probably the best measure of reproductive activity. Sexually active adult males often exhibit specific behavior characteristics that distinguish them from subadults or females. For example, adult male dogs typically lift a hindleg to urinate, a pattern not found in juveniles or females. The scent-marking postures of the male panda are also different from the female [Kleiman, In press (a)]. Since behavioral differences or changes are usually relative rather than absolute, the filling-in of check-sheets for fixed time periods would be of value in the determination of male reproductive condition.

As with females, reproductive inactivity may be due to a variety of factors, both environmental and social. Light cycles and temperature may have a direct effect on gonadal function. Sexual inhibition may occur from social inhibition by more dominant animals in a group or through insufficient stimulation from females. The lack of a secure home base or a territory has been shown to affect reproductive function. Unfortunately, the amount of research on the environmental requirements for adequate male reproduction has been limited. Moreover, almost no consideration has been given to methods of controlling or improving male reproductive activity, either naturally or artificially.

MANAGING SURPLUS STOCK

Difficulties in breeding particular species can be overcome by a concerted effort to improve management of captive stock and through the use of some physiological techniques. However, once a zoo has demonstrated its ability to regularly propagate a species, it is faced with another problem, which, if unsolved, can undo all previous efforts. Decisions concerning the surplusing of excess stock can vitally affect the future of any breeding program. If surplus individuals are scattered by sending them to separate zoos, and captive-born young do not therefore propagate themselves, breeding may terminate with the death of the original animals. Thus, zoos must carefully consider the policies of zoos to which offspring are sent, including exhibition facilities, management procedures, and future breeding plans.

For zoos to produce instead of consume wild animals, reduced competition and increased animal and information exhange are essential. It is not sufficient to have bred a species: What is more important is to be breeding a species and distributing the progeny of that breeding program in such a way that viable captive populations are created and maintained. That the proper management of the results of a propagation program is at least as important as the propagation itself, is a point that has been ignored far too frequently (Perry, Bridgwater, and Horseman, 1972).

REFERENCES

Adams, C. E. 1972. Induction of ovulation and A.I. techniques in the rabbit. Vet. Rec. 91:194–197.

Asdell, S. A. 1964. Patterns of mammalian reproduction. Cornell University Press, Ithaca.

Blatchley, F. R., and B. T. Donovan. 1972. Peripheral plasma progestin levels during anoestrus, oestrus, and pseudo pregnancy and following hypophysectomy in ferrets. J. Reprod. Fertil. 31:331–333.

Bradshaw, G. V. R. 1962. Reproductive cycle of the California leaf-nosed bat, *Macrotus californicus*. Science 136:645–646.

Brambell, M. R., I. W. Rowlands, and J. M. Hime. 1969. An An and Chi Chi. Nature 222:1125–1126.

Bridgwater, D. D. 1972. Saving the lion marmoset. Wild Animal Propagation Trust, Wheeling, West Virginia.

Colby, E. D. 1970. Induced estrus and timed pregnancies in cats. Lab. Anim. Care 20:1075–1080.

Conaway, C. H. 1971. Ecological adaptation and mammalian reproduction. Biol. Reprod. 4:239–247.

Cupps, P. T., L. L. Anderson, and H. H. Cole. 1969. The estrous cycle. Pages 217–250 in H. H. Cole and P. T. Cupps, eds. Reproduction in domestic animals. Academic Press, New York.

Dewsbury, D. A. 1972. Patterns of copulatory behavior in male mammals. Q. Rev. Biol. 47:1–33.

Dukelow, W. R. 1971. Reproductive physiology of primates. Lab. Primate Newsl. 10(2):1–15.

Eisenberg, J. F., and D. E. Isaac. 1963. The reproduction of heteromyid rodents in captivity. J. Mammal. 44:61–67.

Eisenberg, J. F., and D. G. Kleiman. 1972. Olfactory communication in mammals. Ann. Rev. Ecol. Syst. 3:1–32.

Epple, G. 1970. Maintenance, breeding, and development of marmoset monkeys (Callithricidae) in captivity. Folia Primatol. 12:56–76.

Epple, G. 1972. Social behavior of laboratory groups of Saguinus fuscicollis. Pages 50–58 in D. D. Bridgwater, ed. Saving the lion marmoset. Wild Animal Propagation Trust, Wheeling, West Virginia.

Evans, C. S., and R. W. Goy. 1968. Social behaviour and reproductive cycles in captive ring-tailed lemurs (Lemur catta). J. Zool. Lond. 156:181–197.

Foote, R. H. 1969. Physiological aspects of artificial insemination. Pages 313–355 in H. H. Cole and P. T. Cupps, eds. Reproduction in domestic animals. Academic Press, New York.

Francoeur, R. T. 1972. Artificial insemination for species in danger. Oryx 11(5):364–366.

Gould, K. G., E. M. Cline, and W. L. Williams. 1973. Observations on the induction of ovulation and fertilization in vitro in the squirrel monkey (Saimiri sciureus). Fertil. Steril. 24:260–268.

Graham, C. E., D. C. Collins, H. Robinson, and J. R. K. Preedy. 1972. Urinary levels of estrogen and pregnanediol and plasma levels of progesterone during the menstrual cycle of the chimpanzee: Relationship to sexual swelling. Endocrinology 91:13–24.

Hafez, E. S. E., ed. 1968. Reproduction in farm animals. Lea and Febiger, Philadelphia.

Hafez, E. S. E. 1971. Reproductive cycles. Pages 160–204 in E. S. E. Hafez, ed. Comparative reproduction of non-human primates. Charles C Thomas, Springfield, Ill.

Hampton, J. K., B. M. Levy, and D. M. Sweet. 1969. Chorionic gonadotropin excretion during pregnancy in the marmoset. Fed. Proc. 28:367.

Hansel, W., and S. E. Echternkamp. 1972. Control of ovarian function in domestic animals. Am. Zool. 12:225–243.

Harlow, H. 1971. Learning to love. Albion Publishing Co., San Francisco.

Hediger, H. 1965. Environmental factors influencing the reproduction of zoo animals. Pages 319–354 *in* F. A. Beach, ed. Sex and behavior. John Wiley & Sons, New York.

Hendrickx, A. G., and M. L. Houston. 1971. Gestation. Pages 269–301 *in* E. S. E. Hafez, ed. Comparative reproduction of non-human primates. Charles C Thomas, Springfield, Ill.

Hodgen, G. D., M. L. Dufan, K. J. Catt, and W. W. Tullner. 1972. Estrogens, progesterone, and chorionic gonadotropin in pregnant rhesus monkeys. Endocrinology 91:896–900.

Hopper, B. R., and W. W. Tullner. 1969. Relationship of urinary estrogen level to ovulation in the rhesus monkey. Fed. Proc. Abstr. 28:No. 2866.

Jainudeen, M. R., J. F. Eisenberg, and N. Tilakeratne. 1971. Oestrous cycle of the Asiatic elephant, *Elephas maximus,* in captivity. J. Reprod. Fertil. 27:321–328.

Jöchle, W. 1973. Coitus-induced ovulation. Contraception 7:523–564.

Jones, R. C. 1971. Uses of artificial insemination. Nature 229:534–537.

Kleiman, D. G. 1970. Reproduction in the green acouchi, *Myoprocta pratti.* J. Reprod. Fertil. 23:55–60.

Kleiman, D. G. In press (a). Activity rhythms in the giant panda: An example of the use of checksheets for recording behavioral data in zoos. Int. Zool. Yearbk.

Kleiman, D. G. In press (b). Scent-marking in the binturong, *Arctictis binturong.* J. Mammal.

Kleiman, D. G. In press (c). Estrous cycles and behavior of captive tigers. *In* R. Eaton, ed. Proceedings of the Second International Conference on World's Cats.

Kleiman, D. G., and P. A. Racey. 1969. Observations on noctule bats breeding in captivity. Lynx 10:65–77.

Kühme, W. 1973. Zum Problem der Fennekzucht (*Fennecus zerda*) im Kölner Zoo. Z. Kölner Zoo 16:49–58.

Lunenfeld, B., V. Insler, and M. Snyder. 1972. Induction of ovulation by therapeutic agents. Pages 689–703 *in* B. B. Saxena, C. G. Beling, and H. M. Gandy, eds. Gonadotropins. John Wiley & Sons, New York.

Martin, R. D. 1968. Reproduction and ontogeny in tree shrews (*Tupaia belangeri*), with reference to their general behaviour and taxonomic relationships. Z. Tierpsychol. 25:409–495; 505–532.

Ovadia, J., J. W. McArthur, O. W. Smith, and J. Bashir-Farahmand. 1971. An individualized technique for inducing ovulation in the bonnet monkey (*Macaca radiata*). J. Reprod. Fertil. 27:13–24.

Perry, J., D. D. Bridgwater, and D. Horseman. 1972. Captive propagation: A progress report. Zoologica 57(3):109–117.

Polge, C. 1972. Increasing reproductive potential in farm animals. Pages 1–31 *in* C. R. Austin and R. V. Short, ed. Artificial control of reproduction. Cambridge University Press, Cambridge.

Preslock, J. P., S. H. Hampton, and J. K. Hampton. 1973. Cyclic variations of serum progestins and immunoreactive estrogens in marmosets. Endocrinology 92:1096–1101.

Rabb, G. B., J. H. Woolpy, and B. E. Ginsburg. 1967. Social relationships in a group of captive wolves. Am. Zool. 7:305–312.

Rasa, A. 1973. Intra-familial sexual repression in the dwarf mongoose, *Helogale parvula.* Naturwissenschaften 60:303–304.

Rosenblum, L. A. 1968. Some aspects of female reproductive physiology in the squirrel monkey. Pages 147–170 *in* L. A. Rosenblum and R. W. Cooper, ed. The squirrel monkey. Academic Press, New York.

Sadleir, R. M. F. S. 1969. The ecology of reproduction in wild and domestic mammals. Methuen. London.

Schaller, G. B. 1972. The Serengeti lion. University of Chicago Press, Chicago.

Stabenfeldt, G. H., and A. G. Hendrickx. 1973. Progesterone levels in the sooty mangabey (*Cercocebus atys*) during the menstrual cycle, pregnancy, and parturition. J. Med. Primatol. 2:1–10.

Tullner, W. W., and C. W. Gray. 1968. Chorionic gonadotropin excretion during pregnancy in a gorilla. Proc. Soc. Exp. Biol. Med. 128:954–956.

van Wagenen, G. 1968. Induction of ovulation in *Macaca mulatta*. Fertil. Steril. 19:15–29.

Weir, B. J. 1973. The induction of ovulation and oestrus in the chinchilla. J. Reprod. Fertil. 33:61–68.

Weir, B. J. In press. Reproductive patterns in Hystricomorpha. *In* I. W. Rowlands and B. J. Weir, ed. Biology of hystricomorph rodents. Academic Press, London.

White, R. J., C. R. Blaine, and G. A. Blakley. 1973. Detecting ovulation in *Macaca nemestrina* by correlation of vaginal cytology, body temperature, and perineal tumescence with laparoscopy. Phys. Anthrop. 38:189–194.

Wilson, J. R., N. Adler, and B. LeBoeuf. 1965. The effect of intromission frequency on successful pregnancy in the rat. Proc. Natl. Acad. Sci. (USA) 53:1392–1395.

Wright, P. L. 1963. Variations in reproductive cycles in North American mustelids. Pages 77–97 *in* A. C. Enders, ed. Delayed implantation. University of Chicago Press, Chicago.

U. S. SEAL

Veterans Administration Hospital, Minneapolis, Minnesota

D. G. MAKEY

Department of Biochemistry, University of Minnesota, St. Paul

Computer Usage for Total Animal and Endangered Species Inventory Systems: A Specific Proposal

CENSUS AND INVENTORY PROPOSAL

This proposal provides a basis for standardizing the formats used for inventorying zoo animals. The proposed format includes data to be used for a national census, endangered species census, studbook purposes, and inventory purposes.

The coding system utilized by SEAMAK ZOOGAD Systems has been used in this proposal for the taxonomic code number, the zoo location code, and the zoo code for vendor/purchaser names. The Mohawk Zoo of Tulsa system, which appears to be a satisfactory coding system, uses a method acquisition/removal code and source/disposal code. The latter system may well serve as a basis for inventory output and acquisition/ release (A/R) summary reports. A summary of the Mohawk system is given in Table 1.

The proposed system would involve recording of data on sheets by the participating zoo (Figure 1). One sheet would be filled out for each animal and punched onto data cards suitable for computer input (Table 2). For those zoos having access to a computer, it will be possible for them to maintain their own inventory files. However, a copy of these sheets will also be sent to a central computer agency to provide data for the related inventory functions listed below. Those zoos not having access to a computer could have their inventory reports generated by the central agency.

Both authors are connected with the Minnesota Zoological Garden, St. Paul, Minnesota.

TABLE 1 Summary of the Mohawk Zoo Endangered Species Inventory Coding System (as of December 31, 1971)

Status Code	In or Out Code		Death Code	
	1st Character	2nd Character	1st Character	2nd Character
Rare, Endangered	Acquisition Code A birth B trade C purchase D donation, private E donation, TZS F loan to US G capture by staff H recapture (after escape) I retrieval (after theft) J loan by US returned M unknown Removal Code N death O trade P sale Q donation, institutional R donation, private S loan to another T release U escape V theft W loan to US returned	Source–Recipient Code A dealer B zoo C private D institution Disposition of Carcass N incinerated O conventional P given or sent to institution, not OSU Q mounted or preserved R given or sent to OSU S other	Cause of Death A sacrifice B self-inflicted injuries C injury from other agency D malicious destruction E degenerative causes F infectious nonparasitic G parasitic H metabolic I other	Autopsy A yes

TABLE 2 Data Card Format

Parameter	Number of Characters	Card Code
Taxonomic ID	15	A
Individual ID number	6	B
Location (zoo code)	4	C
Sex	1	D
Birth date	6	E
Sire (individual ID number)	6	F
Dam (individual ID number)	6	G
Date in or out	6	H
Acquisition/removal code	1	I
Source/disposal code	1	J
Cause of death	1	K
Purchase/sale price	5	L
Delivery cost	4	M
Vendor/purchaser (zoo code)	4	N
Previous owner ID	6	O
Unused card columns	8	P

Functions of the System

A. Total Species Inventory (a summary of data from all participating zoos)
  1. Species
  2. Sex
  3. Age
  4. Location

B. Endangered Species Inventory and Census
  1. Species
  2. Sex
  3. Age
  4. Location
  5. Births
  6. Deaths
  7. Transfers

C. Studbook Inventories, Census, and Tabulations (same as item B above)

D. Individual Zoo Inventory

E. Annual or Semiannual Acquisition/Release Summary for Individual Zoos (NOTE: In addition to the current A/R Summary, it would be possible to review A/R's for past years.)

F. Family Tree (studbook genealogy)
  1. Determination of an individual's lineage until blanks appear
  2. Summary data on each individual of the tree
    a. Sex
    b. Individual ID number
    c. Location/locations (with dates of acquisition/removal)

   d. Removal code if other than sale (i.e., death, reason)
   e. Birth date
   f. Sire ID number and location
   g. Dam ID number and location
G. Purchase Cost Summary by Species by Year

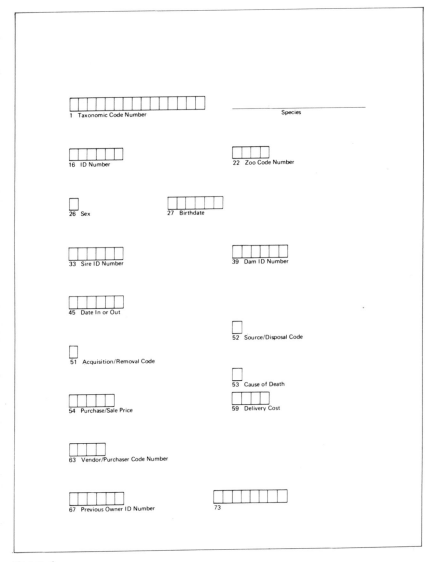

Figure 1
Example of Basic Data Sheet.

Additional Notes

1. Individual Animal ID

   a. It is recommended that numbers be used exclusively for animal identification. Experience with SEAMAK ZOOGAD Systems has demonstrated that spelling of names by people is unreliable identification for a computer. Animal accession number may be a solution to this problem.

   b. For inter-zoo transfers of animals, it is essential to record the animal's previous identification number as well as the previous zoo location code. This establishes a chaining of records, which is essential for the generation of a family tree.

   c. If the proper chaining of records is established as indicated above, the identification of sire and dam need only be recorded on the one data sheet indicating the animal's birth.

2. Vendor/Purchaser Coding

   a. Major animal dealers should be assigned a unique code number. This provision has already been designed into the SEAMAK ZOOGAD ZOO coding system.

   b. Donations of animals from private parties may be indicated by a unique code number, with the name and address of the party being maintained by recipient institution since this is not of general interest.

3. Cause of Death

   The system used by the Mohawk Zoo could be expanded to include autopsy findings as well as cause of death. The numerical system of the *Standard Nomenclature of Veterinary Diseases and Operations,* a 9-digit code, could also be used.

4. Other Data

   Table 3 of this proposal contains a brief review of some studbooks now in effect and the data used by them. This is compared to the proposed system.

   Consideration must be given to other data to be included in the basic system, such as breeding fitness, etc. so that a system of maximum productivity can be made with a minimum of data entry.

COMPUTER PROGRAMMING

The concept guiding the design of the computer programs for the proposed system is centered around program language and file structure. Some well-accepted language, such as COBOL, must be used for programming so that these programs may be easily adapted to other computer facilities. The file structure must be designed to use a minimum of rapid-access storage space since this space is usually at a premium on most computer systems. Thus, the basic programs and file structure suitable for

TABLE 3   Information Collected for Several Studbooks

Type of Information	Lama vicugna (No. 12 p. 147)	European Bison (No. 7 p. 187)	Mongolian Wild Horse	Golden Marmoset	Père David's Deer	Proposed System
Taxonomic code						X
Individual ID	X	X	X	X		X
Sex	X	X	X	X	X	X
Studbook name	X	X	X	X		X
House name	X	X	X	X		
Born	X	X	X	X	X	X
Died	X	X	X	X	X	X
Sire	X	X	X	X		X
Dam	X	X	X	X		X
Location	X	X	X	X	X	X
Breeder		X				
Change of owner		X	X	X		X
List of offspring		X	X			X
List of forebears		X	X			X
Location of birth		X	X			X
Disposal of corpse			X			X
Breeding fitness		X				
Pure blood/hybrid		X				

maintaining zoo animal inventory lists would be adapted to most medium-sized computers. This will allow individual zoos to maintain their own inventory on a computer already maintained by city or state governments.

The system defined in further detail on the following pages requires three file systems and the interpretive programs for input and output of data. The inventory data file contains all the animal inventory data records in the coded form previously illustrated. Due to the coding of taxonomic name and the zoo/dealer names, two files containing the code number/name relationships are needed for the output of these data records into a format which is readily interpreted.

Table 4 shows the mass storage requirements for a nationwide system with immediate access via teletype. It is apparent that a large amount of storage area that is immediately accessible by the computer is required. Utilizing a conservative estimate (1 cent/track/day), the initial costs for the storage space required would be approximately $50 per month and would increase to $150 per month in 10 years. In addition to these costs, the computer required to handle these large amounts of data is approximately $1,000 per CPU hour or $0.28 per CPU second.

The remote site at the zoo would require a teletype to communicate with this computer. The cost of a teletype is approximately $1,000 for

TABLE 4   Computer Capability Statistics for Individual Zoos and for a National Inventory System

Type of System/Parameter	Entries	Characters
Individual zoo (80 column cards)		
Taxonomic code	500–2,000	40 K–160 K
Zoo code	100–1,000	8 K– 80 K
Inventory of entries	500–6,000	40 K–480 K
Initial totals at 20% increase		
in inventory entries	1,100–9,000	88 K–720 K
10-Year inventory	500–6,600	92 K–792 K
National inventory system (40 column cards)		
Initial totals	157,000	6,600 K
10-Year total at 10% increase/yr	471,000	19,480 K
National inventory system (80 column cards)		
Taxonomic codes	6,000	480 K
Zoo codes	2,000	160 K
Mammal inventory	60,000	4,800 K
Bird inventory	65,000	5,200 K
Reptile and amphibian inventory	24,000	1,920 K
Initial totals at 20% increase[a]	157,000	12,560 K
10-Year inventory	467,000	37,360 K
10-Year total	475,000	38,000 K

[a] 20% increase in inventory entries/year = 31,000 entries/year.

purchase and the related transmission devices approximately $500, for a total investment per zoo of approximately $1,500. In addition to these costs, telephone transmission to the computer may be as high as 40 cents per minute for those zoos located remotely to the central computer. At this rate one line of teletype printing costs 4.9 cents.

Totaling the direct costs for remote terminal operation, it is not unreasonable for 50 lines of teletype printing to cost $2.50. One must also note that this figure does not include the cost of the teletype for the individual zoo nor does it include the mass storage costs which must be shared by the participating zoos. Finally, the program development costs previously described have not entered into this figure. Thus, it appears that an "on-line" approach, while technically feasible, is not a practical approach to the problem at hand. Note too that the costs were those of mid-1973 and have not considered inevitable inflationary increases. Estimated start-up costs at this time were $50,000, with total annual operating cost of about $20,000.

DESCRIPTION OF THE PROPOSED SYSTEM

Inventory Program

1. Definition file structure for zoo inventory data, taxonomic code

system, and zoological code system. System must have compatible blocking for disc and tape reads and writes. System should be defined with enough detail for access by assembly language programs if desired.

Programming (30 h)	$1,050
Computer time (1.0 h CPU, 10 h PE)	100
	$1,150

2. Master taxonomic code list—from SEAMAK

a. Reorganization of mammal taxonomic code to include 4-digit subspecies code and a marker between taxonomic name and common name.

b. Insertion of 4-digit subspecies code and marker in taxonomic code system for birds.

c. Construction of taxonomic code system for reptiles and amphibians.

d. Development of program to place the mammalian, avian, reptilian and amphibian codes into a tape structure compatible with the overall program structure.

Programming (10 h)	$ 350
Computer time (0.5 h CPU, 5 h PE)	80
	$ 430

3. Master zoo code list

Direct incorporation of SEAMAK code with only minor changes and additions.

Programming (5 h)	$ 165
Computer time (0.2 h CPU, 2 h PE)	32
	$ 197

4. Taxonomic code/name table

a. Creation of a table containing name and code numbers of all animals contained in zoo inventory data.

b. Table made by scanning master record tape to determine exact species present.

c. Creation of table by extracting appropriate code/names from the taxonomic/zoo code tape.

d. Table must be made accessible by the output program.

Programming (20 h)	$ 700
Computer time (0.5 h CPU, 5 h PE)	70
	$ 770

5. Zoo code/name table

a. Creation of a table containing the code number and name of all zoos or dealers found in the zoo inventory system.

b. Table constructed by scanning master record tape to determine zoo codes used.

c. Extraction of code/names from taxonomic/zoo code tape.

d. Table must be made accessible by the output program.

Programming (10 h)	$ 350
Computer time (0.3 h CPU, 3 h PE)	46
	$ 396

6. Zoo inventory data input

a. Reading of the data inventory cards and placing these records in a format compatible with the recall programs.

b. Storage of these records on the master record tape.

Programming (10 h)	$ 350
Computer time (0.5 h CPU, 5 h PE)	65
	$ 415

7. Individual zoo (monthly) inventory

a. Reading of transaction data records from master tape by zoo onto disc.

b. Sorting of transaction data on basis of taxonomic code and animal sex.

c. Searches made of both taxonomic and zoo code tapes to provide names matching the code numbers found on data records.

d. Writing of headers for classes, orders, etc., of animals.

e. Output of format of data record with interpretation of codes.

f. Total of animals by sex for species, order, class, etc.

Programming (30 h)	$1,050
Computer time (1 h CPU, 10 h PE)	170
	$1,220

8. Total species inventory (endangered species census; studbooks)

a. Scanning of master record tapes with the accumulation of data from individual records pertinent to this report.

b. Sorting of this data by species, sex, and location.

c. Formatted output of the data by species with subdivisions by sex within species.

d. Data output—species, individual ID, sex, age, and location.

e. Individual totals for each species by sex.

Programming (20 h) $ 700
Computer time (0.8 h CPU, 8 h PE)   130
                                                    ─────────
                                                     $ 830

Acquisition/Release Summary
1. Reading of zoo inventory data records from master tape to disc for the year requested.
2. Sorting of inventory data by parameters
    a. Zoo
    b. Date within year
    c. Taxonomic code
3. Formatting and output of data for zoo(s). Format essentially as monthly with exception of date emphasis. Interaction with taxonomic and zoo code/name tables required.

Programming (40 h) $1,400
Computer time (1.5 h CPU, 15 h PE)   245
                                                    ─────────
                                                     $1,645

Ancillary Programs
1. Security copy tapes of records and codes
    a. Provision of security tape copies of all zoo data records, taxonomic code lists, zoo code name lists, and all programs utilized in the total information system.
    b. Developmental cost.

Programming (21 h) $ 735
Computer time (0.4 h CPU, 4 h PE)   70
                                                    ─────────
                                                     $ 805

2. Alter/delete program
    a. Program needed to alter data records which contain erroneous information or delete records which are invalid.
    b. Developmental cost.

Programming (20 h) $ 700
Computer time (0.4 h CPU, 4 h PE)   70
                                                    ─────────
                                                     $ 770

3. Record tabulation by species
    a. For individual zoo.

b. For total inventory (sort by zoo, species within zoo, individual ID, sex, age).

c. Developmental cost for combined programs.

Programming (30 h)	$1,050
Computer time (0.4 h CPU, 4 h PE)	80
	$1,130

4. Card proof reading

a. Program basis: (1) alpha characters in numeric fields; (2) mandatory fields filled, i.e., A, B, C, D, H, I, J, E, F and G if I, K if I, N if I.

b. Developmental cost.

Programming (10 h)	$ 350
Computer time (0.1 h CPU, 1 h PE)	18
	$ 368

c. Estimate of approximately 20,000 cards per hour for proof reading inventory data cards.

5. Zoo proof reading of initial data

a. Formatted listing of zoo data to be entered. Requirements: (1) taxonomic codes and names accessible by program; (2) zoo codes and names accessible by program; (3) all alpha field codes must be accessible by program.

b. Developmental cost.

Programming (15 h)	$ 525
Computer time (0.1 h CPU, 1 h PE)	20
	$ 545

c. Estimate of approximately 10,000 cards per hour can be printed in a format for proof reading.

IV

SPECIAL APPLICATIONS

GEORGE A. SACHER

Division of Biological and Medical Research
Argonne National Laboratory, Argonne, Illinois

Use of Zoo Animals for Research on Longevity and Aging

Biological gerontology, or gerobiology, is the discipline that seeks to comprehend the finitude of individual existence in all its aspects. This definition is broader than the commonly expressed view that gerobiology deals with the "biology of aging," but the broader view is essential for the attainment of sound understanding and effective control of the mechanisms of longevity. Only the acceptance of this more inclusive conception of the scope of gerobiology can open the possibility of real participation in gerobiological research by zoos and zoologists.

One major aspect of the finitude of life is *ontogenetic aging,* the progressive physiological and biochemical deterioration that occurs during the lifetime of the organism. The greater part of the total research effort in gerobiology is directed toward discovering the cellular–molecular basis of the various ontogenetic aging processes, and in fact it is because the absorption with this problem is so complete that gerobiology is so often characterized as the "biology of aging." Two other aspects of the finitude of life must also be understood, however, before man can begin to attain rational and effective procedures for extending his productive lifespan. One of these aspects is the phenomena that govern the vulnerability of the organism to debility, disease, and death. The area of inquiry essentially concerned with understanding the nature of the chance failures of life processes that are responsible for the ever-increasing probability of *qualitative* degradation of organismic function is *thanatobiology*. As such, it is clearly distinguished from research on the ontogenetic aging

This work was supported by U.S. Atomic Energy Commission.

191

process per se, which is primarily concerned with *quantitative* chemical and morphological change. The second aspect of gerobiology in addition to aging per se is the phenomena that govern longevity. A third major subdiscipline seeks to understand the factors responsible for the large differences in longevity that exist between species, and specifically to discover the genetic mechanisms whereby the selection pressures arising from the ecological specialization and reproductive pattern are brought to bear on the temporal dimensions of the species phenotype.

These subdisciplines, dealing respectively with the biology of aging, death, and longevity make up the autonomous discipline of gerobiology. The roles that zoos can perform in aging research are different in kind for each subdiscipline and range widely from utilitarian service to autonomous basic research. I shall do my best to give a comprehensive view of these relationships as they exist at present and as they may possibly develop.

ONTOGENETIC AGING: THE PROVENANCE OF ANIMAL MODELS

The rationale given for the use of animal models for research on aging is that aging is a universal characteristic of living systems, so that all species —at least, all metazoan species—are valid models for the investigation of the basic aging processes. Given this viewpoint, the main considerations in the choice of a model species are those of familiarity, economy, and experimental convenience. It is not surprising, therefore, that the greater part of aging research is carried out with such familiar test objects as mice, rats, and fruit flies. The most important innovation in recent years has been the introduction of lower invertebrate models, such as rotifers, planarians, and nematodes. These organisms have various advantages for research, including short lifespan and simple somatic organization.

The general paradigm of ontogenetic aging research offers little opportunity for the active participation of zoos, but they can nevertheless perform an important role in the development of techniques for successful breeding of new species in captivity. For example, there are several good reasons—low chromosome numbers and short lifespans, etc.—for developing a small marsupial species as a model system for aging studies. Among New World marsupials, the 4-eyed opossum (*Philander opossum*) has promise (Farris, 1950); the Australian marsupials, especially some of the small dasyurid species, also offer a wide field of possibilities (Martin, 1965).

Another physiological characteristic sought by researchers on aging is controllable heterothermy, such as manifested in the diel torpor of pocket mice, in particular the little pocket mouse, *Perognathus longimembris* (Eisenberg, 1967). The availability of such an experimental model would make possible research on the relation of rate of aging and lifespan to body temperature and metabolic rate, a field in which there is at present much speculative interest (Liu and Walford, 1972) but woefully little substantive research (Sacher, 1967). Success in breeding pocket mice and other heteromyid rodents would open the way to interesting new lines of aging research.

There is also strong interest in the development of convenient poikilothermic vertebrate models, such as small fishes or reptiles, with comparatively short lifespans and good reproductive performance. Some research is now being done with the "annual" fishes of the genera *Cynolebias* (Liu and Walford, 1969) and *Oryzias* (Fineman, Hamilton, and Siler, 1974), and more effort in this direction would be justified.

In summary, although most of the requirements for generalized models for ontogenetic aging research can be met with conventional laboratory animals, special circumstances exist where zoos can make a contribution through the adaptation of new species to research environments.

THANATOBIOLOGY: THE CONTINGENT NATURE OF DISEASE AND DEATH

Although a vast amount remains to be learned about aging as a temporal process of loss of quantitative biosynthetic and biophysical capacities, this is nevertheless a well-mapped region in comparison to the aspect of qualitative aging loss, as manifested in the disintegration of organized structure and behavior and in the increased risk of disease and death.

The subdiscipline concerned with these latter questions can appropriately be called *thanatobiology,* the biology of death, because the changes of state from health to disease (or from one disease state to another) have many important features in common with the transition from the living state of death. The major distinction between them—and it is not a fundamental one—is that senescence is the accumulation of a multiplicity of local irreversible changes, "little deaths," arising from microscopic accidents, while organismic death is the consequence of a catastrophic failure.

The primary concern of thanatobiology is to establish the biophysical and biochemical nature of these stochastic transition processes. This is a problem as basic and important as the problem of biochemical aging, for the ability to avert these transitions would be tantamount to the elimination of the corresponding kinds of senescent debility and disease.

This problem is grossly neglected at present because the study of such probabilistic events does not fit well into the dominant paradigm of gerobiology. It also has the disadvantage of lying in the border region between academic aging research and disease-oriented medical research, and research at the interfaces between biology and medicine presents greater than ordinary difficulties of organization, staffing, and funding. Nevertheless, research on age-dependent lesions is going forward and has thereby given rise to a requirement for animal models that are susceptible to a variety of age-related diseases. One example that comes to mind is the research of Drs. Wayland and Ingram at California Institute of Technology on changes in the microcirculation arising from aging and diabetes. This research has focused attention on species that are diabetes-prone in laboratory environments, such as the sand rat (*Psammomys*) and the spiny mouse (*Acomys*).

Another area of age-related disease where there is urgent need for a broader taxonomic base is cancer research. At present there is an overcommitment to the laboratory mouse as the animal model for cancer research. Its ancestor, the house mouse (*Mus musculus*), is at an extreme of evolution toward rapid reproduction and short lifespan, and the various inbred laboratory mouse strains are products of many generations of artificial selection toward even greater fecundity in early life. It is possible that selection toward rapid reproduction is accompanied by the loss or weakening of some genetically controlled suppressors of cancer expression since they have little selective advantage in populations with high reproductive rates and short lifespans. It would, therefore, be prudent to increase our emphasis on the study of longer-lived rodent species with lower cancer incidence rates and rates of aging in order to determine the factors responsible for their lower vulnerability to cancer. Among the species that come to mind, the North American cricetid rodents of the genus *Peromyscus* deserve particular attention. They are easily bred and managed, and they are long-lived and hardy. In my colony at Argonne Laboratory, two *P. leucopus* have lived more than 8 years, and a number of peromyscines of other species are living at over 7 years of age. A beginning has been made at establishing defined populations of *P. leucopus* and *P. californicus* in our laboratory, and *P. maniculatus, P. polionotus,* and the related species *Onychomys leucogaster* are being characterized in other laboratories. Efforts are also under way in several institutions to work out their biochemical genetics and immunogenetics. These efforts need to be strengthened, and zoos and their associated institutes are well fitted to unravel some of the problems in evolutionary genetics.

Despite the concern expressed above about our possible overinvestment in the laboratory mouse, the genus *Mus* is nevertheless an extremely valuable resource for research on the genetics of various oncological, immunological, and neurobiological processes. The full potential of the mouse as a genetic tool cannot be realized, however, until we go back to the wild sources and establish the broadest possible collection of populations. This will maximize the genetic variance of the linkage systems that control the expression of cancer incidence and aging.

Perhaps what I am saying here could be expressed more succinctly as a call for the establishment of rodent research centers dedicated to maximizing the utility of rodents for research on cancer, aging, and other general problems of the mammalian constitution.

EVOLUTIONARY BIOLOGY OF LONGEVITY AND AGING

We come now to the third leg of the gerontological research tripod; it has to do with the evolutionary factors governing animal longevity and the genetic and biochemical mechanisms whereby evolutionary selection pressures are expressed. The evolutionary approach is concerned with discovering the positive, genetically controlled mechanisms on which natural selection can operate to achieve longer life as a basis for the eventual amelioration of aging processes through manipulation of their genetic control systems. Some questions that must be answered at the outset are as follows:

• Does the longevity of mammalian species have consistent constitutional correlates?
• What is the relation of longevity to other life history parameters, especially the parameters of development and reproduction?
• What are the selective forces that have operated to bring about the large range of mammalian lifespans?
• Is there a single set of longevity-controlling mechanisms held in common by all mammals, or are there diverse mechanisms in the various phyletic groups?

Without the answers to these and related questions, we cannot develop an optimum research program for increasing man's vigor and longevity, or even make wise choices of animal models.

My own concern has been to discover how lifespan is related to other constitutional dimensions [Sacher, 1959, In press (a) and (b)], and I can summarize my findings by saying that lifespan is found to be significantly related to four constitutional variables. Table 1 gives the multiple regression relation of logarithm of lifespan to these four variables:

TABLE 1 Multivariate Allometric Relation[a] of Species Lifespan (L) to Adult Brain Weight (E), Body Weight (S), Resting Metabolic Rate per Gram of Body Weight (M), and Body Temperature (T)[b]

$$\log \text{(lifespan)} = (0.62 \pm 0.07)\ \log E$$
$$- (0.41 \pm 0.05)\ \log S$$
$$- (0.52 \pm 0.07)\ \log M$$
$$+ (0.026 \pm 0.009)\ T$$
$$+ 0.89 \pm 0.32$$

[a] Computed as a linear least squares regression of logarithm of L on logarithms of E, S and M, and on T, based on 85 mammalian species sampled from all major orders except bats.
[b] E and S in grams, M in calories per gram hour, T in degrees Celsius.

body weight, brain weight, metabolic rate, and body temperature. This multivariate relation has a squared multiple correlation of 0.72, which means that 72 percent of the lifespan variance of mammals is accounted for by these four variables. This is an important result, for it disproves the prevalent notion that species can evolve different lifespans in arbitrary ways. All future work on mammalian longevity must start from the fact of its close relationship to these constitutional variables.

Research on the phylogeny of longevity can be no better than the lifespan and pathology records on which it is based. It is fortunate, therefore, that in each generation there has been a compiler whose complete devotion to his work is matched by his extensive knowledge and his high critical standards. The work of P. Chalmers Mitchell early in the century and of Major Stanley Flower in the 1920's (Flower, 1931) is now being carried on with greatly increased scope and detail by Marvin Jones (Jones, 1968), whose exhaustive tabulation of data from the zoos of the world on the longevity and reproduction of mammals in captivity has been an invaluable resource for my research on the comparative biology of longevity in mammals. The continued accumulation of such data is a vital and unique responsibility, and the movement toward uniform recording and reporting that has been discussed in this symposium (Seal, this volume) should be encouraged; such data are valuable for research in gerobiology and other fields, as well as for the improved management and preservation of captive animal populations.

When adequate vital statistics on zoo populations become available, it will be possible to apply actuarial techniques to the population biology of captive animals and thus redress the unfortunate situation that exists at present in which more can be said about the life tables of animals in the wild than in captivity.

Although virtually all the lifespans, and a major proportion of the brain weight and body weight data used in my analysis came from zoo materials, the metabolic rates and body temperatures came almost entirely from university laboratories. There are still major questions to be answered about the phylogeny of energy metabolism patterns and about acclimation to artificial environments. Much of this information will be useful for comparative aging research because, for many purposes, the lifetime energy expenditure per gram of tissue is a more rational measure of lifetime performance than the number of years lived. Also, there are major gaps in zoogeographical coverage that zoos could fill at some advantage to themselves. A particularly neglected topic is the physiological ecology of the caviamorph rodents. We have virtually no data on the thermoregulation and metabolism of this major taxon; a zoo specializing in them could begin to explore this important physiological problem.

CONCLUSION

Some of the research directions discussed here are far removed from conventional views of the scope of aging research. The justification for including them derives from the broader conception of gerobiology outlined in the opening paragraphs. There will inevitably be an increased realization that the problems of aging and cancer demand a broader understanding of the factors responsible for the diversity of mammalian life patterns, at all levels from the molecular to the systemic, including the role of population biology and evolutionary factors. When that day arrives, zoos will at last come into their own as centers of research.

REFERENCES

Eisenberg, J. F. 1967. The heteromyid rodents. Pages 391–395 in The UFAW handbook. Universities Federation for Animal Welfare, London.

Farris, E. J. 1950. The opossum. In E. J. Farris, ed. The care and breeding of laboratory animals. John Wiley & Sons, New York.

Fineman, R., J. Hamilton, and W. Siler. 1974. Duration of life and mortality rates in male and female phenotypes in three sex chromosomal genotypes (XX, XY, YY) and in the killifish, Oryzias latipes. J. Exp. Zool. 188:35–40.

Flower, S. S. 1931. Contributions to our knowledge of the duration of life in vertebrate animals. Pages 145–234 in V. Mammals. Zool. Soc. London Proc.

Jones, M. L. 1968. Longevity of primates in captivity. Int. Zool. Yearbk. 8:183–192.

Liu, R. K., and R. L. Walford. 1969. Laboratory studies on lifespan, growth, aging and pathology of the annual fish, Cynolebias bellottii Steindachner. Zoologica 54:1–16.

Liu, R. K., and R. L. Walford. 1972. The effect of lowered body temperature on lifespan and immune and non-immune processes. Gerontologia 18:363–388.

Martin, P. G. 1965. The potentialities of the fat-tailed marsupial mouse, Sminthopsis crassicaudata (Gould), as a laboratory animal. Austral. J. Zool. 13:559–562.

Sacher, G. A. 1959. Relation of lifespan to brain weight and body weight in mammals. Pages 115–133 in G. E. W. Wolstenholme and M. O'Connor, ed. CIBA Foundation Colloquia on Aging, Churchill, London.

Sacher, G. A. 1967. The complementarity of entropy terms for the temperature-dependence of development and aging. Ann. N.Y. Acad. Sci. 138:680–712.

Sacher, G. A. In press (a). Maturation and longevity in relation to cranial capacity in hominid evolution. *In* R. Tuttle, ed. Antecedents of man and after. I. Primates: Functional morphology and evolution. Mouton & Co., The Hague.

Sacher, G. A., and E. F. Staffeldt. In press (b). Relation of gestation time to brain weight for placental mammals: Implications for the theory of vertebrate growth. Am. Naturalist.

ROBERT M. SAUER, V.M.D.

National Zoological Park
Smithsonian Institution, Washington, D.C.

Lead Poisoning; Selenium-Vitamin E; Sudden Infant Death; Ischemic Heart Disease: Zoological Research of Comparative Medical Interest

The following material concerns two research projects currently being conducted at the National Zoological Park that have implications which go beyond the boundaries of the Zoo.

LEAD POISONING

Lead intoxication has been diagnosed in 42 primates, 11 parrots, and 3 fruit bats at the National Zoological Park (Zook, 1971, 1973; Zook, Sauer, and Garner, 1970, 1972a, Zook et al., 1973). Diagnoses were made clinically either by the observation of signs of encephalopathy and the finding of 200 milligrams of lead or more per 100 milliliters of blood, or postmortem by the presence of renal acid-fast intranuclear inclusion bodies and excess lead in liver specimens. Sources of lead exposure were investigated. Samples of cage paint, diets, drinking water, air and atmospheric fallout were analyzed for lead content. Significant sources of lead were essentially found only in lead-containing paint. The lead content of food, water, and air was similar to that available to most urban Americans which to date has not been shown to cause lead poisoning.

Of particular interest were the 42 primate cases, 25 of which had signs or lesions of lead encephalopathy (Sauer and Zook, 1970; Zook, Sauer, and Garner, 1972b; Zook and Sauer, 1973). Although all monkeys had essentially similar exposure to lead sources, the occurrence of lead poisoning was predominantly in the subfamily Cercopithecinae and juveniles were affected more frequently than adults. The occurrence of encephalopathy was especially high in juvenile Cercopithecinae and was evident in as short a period of time as 6 weeks following initial exposure to lead painted cages. Reports in the literature over the past 50 years confirm the high incidence of lead poisoning in Old World primates.

Zook, Eisenberg, and McLanahan (1973) explored the factors underlying these observed species' differences. They found evidence that the foraging and grooming behavior of the Cercopithecinae may predispose them to ingestion of lead-containing cage paint.

Allogrooming is high among the Cercopithecinae and, while it is not believed to be a major source of lead poisoning, paint chips and atmospheric fallout on the pelage of a partner could represent a contribution.

Many of the Cercopithecinae who are adapted to foraging on the ground, on bark, or on rotten limbs also do much artifact manipulation. Thus, when confined to painted cages, their feeding habits predispose them to pick, gnaw, and ingest paint from bars and other artifacts. Lead-containing paint has been evidenced by radiographs and postmortem in the digestive tracts of these animals. Stereotyped gnawing and licking behavior can also be provoked by social stress, and the subordinate animal is most prone to this form of behavior.

Juveniles are in double jeopardy of lead poisoning. They hold a low social position and have a tendency to gnaw and mouth objects when teething.

We now have evidence that the brain damage so commonly associated with lead poisoning of zoo-dwelling primates is not necessarily dose-related. Other factors, environmental, social, or dietary, must play a role. Studies are presently under way to explore the impact of these other factors in nonhuman primates.

SELENIUM–VITAMIN E DEFICIENCY

During the past several years at the National Zoological Park (NZP) many seemingly unrelated problems involving three classes of animals—reptiles, birds, and mammals—have emerged as a syndrome of signs related to Se–vitamin E (Se–E) deficiency (Sauer and Zook, 1972).

Similar syndromes have been recognized for years in the domestic food-producing animals, sheep, cattle, and poultry, and in the horse as well. Histologic lesions of Se–E myopathy can be found in a variety of laboratory animals used in routine nonnutritional studies.

To ascertain if a dietary deficiency existed at NZP, forages and 20 commercial rations, including milk substitutes for humans and animals, were analyzed for Se–E content. All forages (Table 1) were deficient in minimum requirements for Se (<0.1 ppm) (Alloway, 1968). Com-

TABLE 1   Selenium Content of Forages at the National Zoological Park

Forage Type	Selenium, ppm
Timothy hay	ND[a]
Prairie hay	ND
Alfalfa hay (eastern)	ND
Alfalfa hay (South Dakota)	0.18
Silage	ND
Bamboo (Site 1)	ND
Bamboo (Site 2)	ND
Bamboo (Site 3)	ND
Bamboo (Site 4)	ND
Bamboo (Site 5)	ND
Rolled oats	ND

[a] ND = None Detected. Recovery was 96.6 ± 12.9% for amounts of 1, 2, and 3 µg selenium standard analyzed.

mercial pelleted feeds and supplements were variable (Table 2). Some were more than adequate if fed as a complete diet; others were markedly deficient. When complete diets were analyzed for the proportions of high and low Se components, all ungulate diets and many avian diets were considered inadequate. Monkey, dog, and rat and mouse chows varied

TABLE 2   Selenium Content of Commercially Compounded Feeds

Feed Type	High Selenium, ppm	Low Selenium, ppm
Sweetfeed	0.18	0.12
Horse pellets	0.13	0.10
Monkey chow (brand 1-A)	0.23	0.22
Monkey chow (brand 1-B)	0.22	0.16
Monkey chow (brand 1-C)	0.12	0.09
Monkey diet	0.11	0.06
Game bird feeder layena	0.10	0.06
Game bird startena	0.10	0.08
Turkey finishing chowder	0.56	0.48
Trout chow	0.42	0.36
Laboratory chow (brand 1)	0.23	0.20
Laboratory chow (brand 2)	0.13	0.10
Dog chow	0.10	0.05
Dog meal	0.18	0.17
Rabbit chow	0.20	0.19
Guinea pig chow	0.41	0.32
Mouse chow	0.09	0.09
Rat chow	0.11	0.04
Deer pellets (experimental)	0.29	0.28
Deer pellets (brand 1)	0.20	0.18
Deer pellets (brand 2)	0.07	—

from highly deficient to inadequate in Se content. While all commercial feeds met the supposedly minimum standards for vitamin E, most were fortified with α-tocopherol (Table 3).

It has been demonstrated that selenium and vitamin E are vital components of biologic membranes and enzyme systems (Gruger and Tappel, 1970; Lucy, 1972; Rotruck et al., 1973; Smith, Tappel, and Chew, 1974). The evidence continues to mount. Because humans share the same environment, the same foods, the same biologic membranes and the same enzyme systems as our exotic, laboratory, and domestic animals, it is axiomatic to consider that humans and their offspring might also share the same problems.

Of all the Se–E deficiency syndromes seen at the National Zoological Park, the one of most comparative medical interest is that of skeletal and cardiac myopathy. This syndrome affects animals that are stressed particularly during handling or shipping, when it is referred to as "capture myopathy." There is a high incidence in ungulate neonates and subadults occasionally resulting in sudden deaths. The condition is akin to white muscle disease of cattle and sheep, but often so acute that muscle lesions may not be observed either at postmortem or histologically. A similar diagnostic problem exists in humans who succumb in minutes to a

TABLE 3 Tocopherol Content of Commercially Compounded Dry Feeds

Feed Type	High Tocopherol, μg/g	Low Tocopherol, μg/g
Sweetfeed	83	78
Horse pellets	440	430
Monkey chow (brand 1-A)	177	170
Monkey chow (brand 1-B)	216	209
Monkey diet	280	277
Game bird feeder layena	153	145
Game bird startena	129	116
Turkey finishing chowder	139	133
Trout chow	224	216
Laboratory chow (brand 1)	141	141
Laboratory chow (brand 2)	156	148
Dog chow	142	137
Dog meal	230	230
Rabbit chow	110	95
Guinea pig chow	60	60
Mouse chow	280	265
Rat chow	163	160
Deer pellets (experimental)	300	300
Deer pellets (brand 1)	156	146

coronary attack since neither gross nor histologic lesions can be seen by routine procedures.

In 1971, a new, nonenzymatic histochemical technique, the hematoxylin-basic fuchsin-picric acid (HBFP) stain was introduced at the Mayo Clinic (Lie et al., 1971). It provided a vivid tinctorial demonstration of early myocardial degeneration. It stained abnormal myocardium bright red, whereas normal or infarcted myocardium stained pale yellow. The results were reproducible and unaffected by postmortem autolysis. The HBFP stain was instituted as a routine histologic procedure at NZP and correlated with serum enzymes designed to detect acute muscle degeneration. Subsequently, the number of diagnoses of myopathy and cardiomyopathy increased exponentially.

This cardiomyopathy syndrome of ungulates and other species is reminiscent of two distinct human diseases which I feel have common denominators in basic Se–E nutrition and a vulnerable myocardium. These two human afflictions are the sudden infant death syndrome, or "crib death," and ischemic heart disease.

SUDDEN INFANT DEATH SYNDROME

Since the suggestion by Money (1970) that the Se–E responsive neonatal deaths of pigs might be analogous to sudden infant death syndrome (SIDS), little has been done to determine the Se–E status of infants. The infant, however, can be no more adequate than the food he or she eats. Table 4 indicates the Se content of a variety of milk substitutes. If one accepts the animal standard of 0.10 ppm as an acceptable nutritional level in the overall diet, then these substitutes, which comprise a large proportion of an infant's diet, are precariously marginal. Whole milk, skim milk, evaporated milk, and strained baby foods—with the exception of meat—offer little support (Hadjimarkos, 1966; Morris and Levander, 1970), and it is a basic concept that a whole cannot be greater than the sum of its parts.

As regards vitamin E, it has been demonstrated that milk, milk products, and simulated milks are low in total tocopherols (Herting and Drury, 1969). Probably of greater significance is a co-existing low or marginal vitamin E:polyunsaturated fatty acid (PUFA) ratio. PUFA content of the diet is the principal determinant of the need for vitamin E and becomes critical with infant formulas containing vegetable oils.

The proof of feeding is the level a nutrient attains in the body. Various studies agree that blood Se–E levels of infants are significantly less than that of adults (Nitowsky, Cornblath, and Gordon, 1956; Fomon, 1967; Rhead et al., 1972 a,b). Such low levels occur at a time in life when

TABLE 4 Selenium Content of Milk Substitutes and Milk Concentrates

Milk Substitute Type	Selenium, ppm
Enfamil	0.02
Enfamil with iron	0.05
Enfamil (add water)	0.06
Enfamil with iron (powder)	0.05
Enfamil (powder)	0.06
Similac with iron	0.02
Similac advance	0.02
Similac—isomil	0.04
Soyalac	0.02
Soyalac single strength	0.04
SNA concentrate with iron	0.02
Neo-mull-soy	0.08
Mull-soy concentrate	0.07
Nutramagen (powder)	0.07
Pro So Bee	0.07
Calfnip	0.08
Esbilac	0.03
KMR	0.09
Medicated lamb milk replacer	0.15
Evaporated milk (Carnation)	0.06
Evaporated milk (Eaglebrand)	0.03
Evaporated milk (Giant Food)	0.05
Evaporated milk (Pet)	0.02

muscle fibers are most vulnerable to PUFA and the balance of other essential nutrients (Bird and Szabo, 1964). Another factor to be considered is that local tissue deficiency of vitamin E can occur and cause death even though no absolute deficiency exists in the body as a whole.

The point to be made clear is that the evidence indicates that both infant foods and infants are precariously marginal, if not absolutely deficient in Se–E. The result is a myocardium vulnerable to a variety of stresses. It is also interesting to note that Ontario, Canada, a province notoriously Se-deficient, has the world's highest incidence of SIDS.

ISCHEMIC HEART DISEASE

Much that has been said concerning the Se–E status of infants can be restated as regards the Se–E status of adults. The Se content of many foods has been published (Morris and Levander, 1970). Adequacy in a diet, however, is dependent on a number of variables, such as (1) the proportions of high and low selenium foods consumed; (2) the degree to which foods are cooked, as many Se compounds are quite volatile; and

(3) the vitamin E content of the diet. Experimentally, the need for Se is, to a degree, inversely related to the availability of vitamin E. Concerning the latter, a recent comprehensive analysis of vitamin E in the American diet concluded that "it is apparent that the types of average U.S. diets included do not meet the recommended allowance for vitamin E" (Bier and Evarts, 1973). Instead of recommending supplementation of the diet the authors suggested that "in the absence of clinical or biochemical evidence indicating that the American public is not receiving sufficient vitamin E, the present recommended allowance is unrealistically high." Such a conclusion defies the overwhelming amount of evidence obtained from dozens of species of animals. It also presupposes that the ultimate in technology now exists and overestimates the diagnostic acumen of the medical profession.

There are those, however, who are beginning to take a second look at the theories of ischemic heart disease that have been in vogue for the past decade or two (Anderson, 1973 a,b). They are considering the fact that maybe a chronically sick myocardium is basic to ischemic heart disease and more vulnerable to atherosclerosis, smoking, lack of exercise, stress, etc.

Recent studies in Sweden (Erhardt, Lundman, and Mellstedt, 1973) have clearly demonstrated that coronary thrombosis is a result rather than a cause of myocardial infarction. Workers in Toronto have suggested that focal myocardial fibrosis in humans may be analogous to similar lesions in sublethal Se–E nutritional muscular dystrophy of animals. It is also pointed out that with the cholesterol fad of the past decade the E/PUFA ratio of American diets has been declining to a critical level known to cause muscular dystrophy (Anderson, 1973 c,d). Further evidence of a role for Se–E in ischemic heart disease comes from a recently completed clinical trial in Mexico on patients with recurring attacks of angina pectoris with or without associated myocardial infarct (Anon., 1972). The drug used was a selenium–tocopherol combination similar to that approved by the FDA for use in animals in 1962. The trials demonstrated beneficial responses in 92 percent of the cases. Improvement was noted in one or more of three areas: (1) reduction or elimination of anginal attacks and nitroglycerin dependence; (2) increased vigor, activity, and work capacity; and (3) improved electrocardiograms.

With the exception of the foregoing example, Se or Se–E combinations have not been tried in humans, and there are those who are cynical about the use of vitamin E (Olson, 1973). Trials of vitamin E in humans have contained two fallacies. First, they have been "after the fact" courses of therapy with pharmacologic doses of vitamin E. This does not take into account the fact that many of the deficiency lesions, as seen in animals, are irreversible. Such "crash" courses of treatment can probably do no more than strengthen residual healthy tissue. Secondly, because the individual and collective roles of Se–E have not been clearly defined, particularly in humans, they should be used in combination.

In conclusion, I feel that the benefits to be derived from Se–E are those of prevention of cardiomyopathy rather than cure. Nutritional levels should begin prenatally and be maintained in the diet throughout life. It is becoming clearer at NZP that beneficial effects from food supplementation in ungulates may take as long as two breeding cycles (2 years) or more to become obvious. Critical evaluation of the prophylactic beneficial effects in man may necessarily take a long time. Short-term judgments should be reserved. Se–E should be used in combination. Vitamin preparations and foods should be supplemented with E as is now the case with A and D. "Selenophobia" must be overcome and the Delaney Act (Shapiro, 1972; Harr *et al.*, 1973) re-examined as concerns the nutritional aspects of Se.

REFERENCES

Alloway, W. H. 1968. Selenium—Vital but toxic needle in the haystack in science for better living. The Yearbook of Agriculture. U.S. Government Printing Office, Washington, D.C.

Anderson, T. W. 1973a. The changing pattern of ischemic heart disease. Can. Med. Assoc. J. 108:1500–1504.

Anderson, T. W. 1973b. Mortality from ischemic heart disease. J. Am. Med. Assoc. 224(3): 336–338.

Anderson, T. W. 1973c. Nutritional muscular dystrophy and human myocardial infarction. Lancet, pp. 298–302.

Anderson, T. W. 1973d. The vulnerable myocardium. Lancet, pp. 1084–1085.

Anonymous. 1972. Medical benefits—Beast to man. J. Am. Vet. Med. Assoc. 162: 906.

Bier, J. G., and R. P. Evarts. 1973. Tocopherols and fatty acids in American diets. J. Am. Dietet. Assoc. 62:147–151.

Bird, W. C., and A. B. Szabo. 1964. Lipid peroxidation in nutritional muscular dystrophy. Proc. Soc. Exp. Biol. Med. 117:345–350.

Boler, J. B. 1969. Deficiency of Q_{10} in the rabbit. Int. J. Vit. Res. 39:281–288.

Erhardt, L. R., T. Lundman, and H. Mellstedt. 1973. Incorporation of ^{125}I-labelled fibrinogen into coronary arterial thrombi in acute myocardial infarction in man. Lancet, pp. 387–390.

Fomon, S. J. 1967. Infant nutrition. W. B. Saunders Co., Philadelphia.

Gruger, E. H., Jr., and A. L. Tappel. 1970. Reactions of biological antioxidants. III. Composition of biological membranes. Lipids 6(2):147–148.

Hadjimarkos, D. M. 1966. Selenium concentration in evaporated milk. J. Pediatr. 68(3):470–472.

Harr, J. R., J. H. Exon, P. H. Weswig, and P. D. Whanger. 1973. Relationship of dietary selenium concentration; chemical cancer induction; and tissue concentration of selenium in rats. Clin. Toxicol. 6(3):487–495.

Herting, D. C., and E. E. Drury. 1969. Vitamin E content of milk, milk products, and simulated milks: Relevance to infant nutrition. Am. J. Clin. Nutr. 22(2): 147–155.

Lie, J. T., K. E. Holley, W. R. Kampa, and J. L. Titus. 1971. New histochemical method for morphologic diagnosis of early stages of myocardial ischemia. Mayo Clin. Proc. 46(5):319–327.

Lucy, J. A. 1972. Functional and structural aspects of biological membranes: A suggested structural role for vitamin E in the control of membrane permeability and stability. Ann. N.Y. Acad. Sci. 203:4–10.

Money, D. F. L. 1970. Vitamin E and selenium deficiencies and their possible aetiologic role in the sudden death of infants syndrome. N. Zeal. Med. J. 71(32): 32–34.

Morris, V. C., and O. A. Levander. 1970. Selenium content of foods. J. Nutr. 100(12):1383–1388.

Nitowsky, H. M., M. Cornblath, and H. H. Gordon. 1956. Studies of tocopherol deficiency in infants and children. II. Plasma tocopherol and erythrocyte hemolysis in hydrogen peroxide. AMA J. Dis. Child. 92:164–174.

Olson, R. E. 1973. Vitamin E and its relation to heart disease. Circulation 48: 179–184.

Rhead, W. J., S. L. Saltzstein, E. E. Cary and W. H. Alloway. 1972a. Vitamin E, selenium, and the sudden infant death syndrome. J. Pediatr. 81:2.

Rhead, W. J., E. E. Cary, W. H. Alloway, S. L. Saltzstein, and G. N. Schrauzer. 1972b. The Vitamin E and selenium status of infants and the sudden infant death syndrome. Bio-organ. Chem. 1:289–294.

Rotruck, J. T., A. L. Pope, H. E. Ganther, A. B. Swanson, D. G. Hafeman, and W. G. Hockstra. 1973. Selenium: Biochemical role as a component of glutathione peroxidase. Science 179(4073):588–590.

Sauer, R. M., and and B. C. Zook. 1970. Demyelinating encephalomyelopathy associated with lead poisoning in nonhuman primates. Science 169:1091–1093.

Sauer, R. M., and B. C. Zook. 1972. Selenium-vitamin E deficiency at the National Zoological Park. J. Zoo Anim. Med. 3:34–36.

Shapiro, J. R. 1972. Selenium and carcinogenesis: A review. Ann. N.Y. Acad. Sci. 192:215–219.

Silver, M. D., T. W. Anderson, A. A. van Dreumel, and R. E. C. Hutson. 1973. Nutritional muscular dystrophy and human myocardial infarction. Lancet. pp. 912–913.

Smith, P. J., A. L. Tappel, and C. K. Chew. 1974. Glutathione peroxidase activity as a function of dietary selenomethionine. Nature 247:392–393.

Zook, B. C. 1971. An animal model for human disease. Lead poisoning in nonhuman primates. Comp. Pathol. Bull. 3(1):3–4.

Zook, B. C. 1973. Lead poisoning in urban pet and zoo animals. Clin. Toxicol. Bull. 3:91–100.

Zook, B. C., J. F. Eisenberg, and E. McLanahan. 1973. Some factors affecting the occurrence of lead poisoning in captive primates. J. Med. Primatol. 62:206–217.

Zook, B. C., and R. M. Sauer. 1973. Leucoencephalomyelosis in nonhuman primates associated with lead poisoning. J. Wildl. Dis. 9:61–63.

Zook, B. C., R. M. Sauer, M. Bush, and C. W. Gray. 1973. Lead poisoning in zoo-dwelling primates. Am. J. Phys. Anthrop. 38(2):415–424.

Zook, B. C., R. M. Sauer, and F. M. Garner. 1970. Lead poisoning in Australian fruit bats (Pteropus poliocephalus). J. Am. Vet. Med. Assoc. 157(5):691–694.

Zook, B. C., R. M. Sauer, and F. M. Garner. 1972a. Lead poisoning in captive wild animals J. Wildl. Dis. 8:264–272.

Zook, B. C., R. M. Sauer, and F. M. Garner. 1972b. Acute amaurotic epilepsy caused by lead poisoning in nonhuman primates. J. Am. Vet. Med. Assoc. 161(6): 683–686.

Summary and Prospect

This concludes our first such comprehensive symposium within an AAZPA annual conference. From comments that I have heard during the past few days, it has been very well received. It has given us an opportunity to bring people from around the United States together both from our staffs at the zoos and aquariums and from some of the universities and research institutions. On behalf of all of the members of AAZPA, I want to personally commend our two overseas guests, Drs. Goodwin and Van den bergh for making time to come and present their thoughts and information to us. Again, this is a first for AAZPA. We want to express thanks for the close support and cooperation of the Institute of Laboratory Animal Resources (ILAR), National Research Council. I believe it is most fitting that the person to really wrap up these two days of the symposium for us is Dr. Cluff Hopla, Department of Zoology, University of Oklahoma, Chairman of the Institute of Laboratory Animal Resources, Division of Biological Sciences, Assembly of Life Sciences, National Research Council.

LESTER FISHER, *President*
American Association of Zoological
Parks and Aquariums

DR. HOPLA'S STATEMENT

I have been delighted to have had the opportunity to visit with many of you and to become better acquainted with your problems and philosophies. Time has passed quickly because the discussions have been substantive and the audience responsive. On behalf of ILAR I wish to thank the members of the original *ad hoc* committee, John Eisenberg, Lester Fisher, George Rabb, and Robert Snyder, who helped to shape the broad outlines of this program.

Some of you undoubtedly wonder what ILAR represents. As Dr. Fisher has indicated, the Institute of Laboratory Animal Resources is a part of the Assembly of Life Sciences and within the Division of Biological Sciences, National Research Council, National Academy of Sciences. Time is of the essence so I will only mention that ILAR is concerned with the welfare of animals used in biomedical research, beyond the provisions of the Animal Welfare Act of 1970, and we are highly interested in animal models and genetic stocks. ILAR has a deep and abiding interest in the conservation of animals, especially nonhuman primates, which gives us a common interest with you. Communications between your membership and individuals within the biomedical research community should be enhanced. It is our hope that this symposium will foster this communication.

ILAR and the Society for Experimental Pathology had a very interesting symposium on marine invertebrate animals as possible models in biomedical research at the FASEB meetings this past spring. The proceedings are now in print. This symposium has been well received. It is highly likely that a proposed guide on the Collection and Care of Marine Invertebrates will be of interest to several of you.

In reviewing the activities of the past few days I would like to touch on something that was outside of the symposium. This was the presentation you had on the federal regulations, chaired by Dr. King. You are not the only people who are confronted with this kind of problem. The whole biomedical research community is involved with them. I was somewhat surprised that Dr. Fisher was the only one who spoke out. I suspect that with his eloquence and succinctness you probably felt no one needed to say more. One very important point was made, that the law and the resultant regulations are not one and the same. You must speak out to help shape the regulations so that they are compatible with the worthwhile goals you have established. If you do not, someone else will because the people in the bureaucratic agencies are subject to political and emotional opinions that result in a different kind of regulation. The federal agencies need your support.

Now to research. Why do zoological parks and aquaria undertake research? *First,* research is undertaken on behalf of the animals because you are committed to improving their health, appearance, reproduction, and longevity. The animals represent a tremendous financial investment and some of the animals that have been relatively easy to obtain in the past will be difficult to procure in the future. *Second,* basic research ultimately benefits man and other animals. It goes beyond the parameters of zoos. *Third,* Père David's deer is an example of an animal living in zoological parks but not in the wild. Ten to fifteen years from now there will be other animals in this category. Destruction of habitat by the ever expanding human population precludes that it will be otherwise. Zoological parks and aquariums have a unique responsibility to perpetuate the existence of rare and endangered animals. *Fourth* is the discovery of new models for biomedical research. Zoological parks have an expertise in knowledge of the handling of exotic animals that we find almost nowhere else in our biological spectrum. You have a great deal to offer.

It is important that large and small zoos undertake research. We have had a good representation of philosophy from this standpoint. The refreshing point here is the cooperation with centers of higher learning in undertaking to develop a sense of awareness of animals in young people. We have discussed basic and applied research and that they do blend. We mentioned something of how research is done—the integration with universities and other research institutions within the vicinity of the zoo, the use of students, starting from the high school level through undergraduate, graduate and postdoctorate, as well as having a core of personnel within the zoological park itself.

What kinds of research can be undertaken? Here I'm speaking mostly of the areas.

Conservation—Zoological parks have had a long history in this area but in some ways you must reach out more broadly in the future. Some animals that we think are plentiful, won't be as the human population increases and competition for protein and other food substances increases. Mr. Conway mentioned the situation with regard to the pipeline in Alaska. I hope you know the Arctic Wildlife Reserve is now opened for a gas pipeline route. In 1965 when this refuge was first set up and Olaus Murrie made his presentations at the Alaska Science Conference it was thought an area had been selected that would be protected from "industrial progress" and exploitation. It now seems there is no place on this planet that is safe. Monday you referred to another example of this regarding the tuna industry and dolphins.

Behavioral research is compatible with the other goals of zoological parks, as are biochemistry, reproductive physiology, nutrition, and comparative pathology, which have all been represented at this conference. In a subtle way, so has systematics and it is this last that I will dwell on just briefly. I am reminded of a workshop on invertebrate models in biomedical research that I attended some months ago. If you weren't a molecular biologist you were on the outer fringe. One of the very exciting models with invertebrates is a barnacle that has giant muscle fibers, likely the largest of any animal in the world. The investigators weren't getting very good results and a systematist happened to look over their shoulders and discovered that they were working with three separate species. He informed the director of the laboratory and the director said, "Why, I presume they were delighted when you told them." A gleeful expression spread across the systematist's face and he said, "I didn't tell them."

There is a place for systematists in zoos but I do not mean to imply that every zoo should have one. Dr. Benirschke's report about the subspecies of squirrel monkeys pointed out realistically that we need to know what we are working with.

Comments were made about a critical mass of researchers being available to have an interchange, to be synergistic. We heard about core support for research programs within zoos because, without it, they are fraught with difficulties and likely will not survive long.

The research staff must recruit funds elsewhere to augment the core support.

Researchers must have freedom to work on a variety of problems and not just on zoo animals. They are there to form a pool of expertise that is available when it is needed. If they are not given this freedom they will not be as innovative as they could be, and eventually many of them will be drawn off elsewhere, where they have this kind of freedom. I have seen this happen with some zoological parks in the past. The researchers were not core-supported; eventually an ultimatum was issued that no research could be done unless it was directly concerned with zoo animals. Within 6 months to a year whole programs had vanished.

Another important consideration is selecting the right talent. Imaginative researchers must be selected who can hold their own with those in other areas of biological and biomedical research. They must compete for grant support to enrich the core support that should form the foundation of any sound research program.

Of course, no research can be undertaken in a zoo if the director is hostile to it. Fortunately, there has been a subtle change or turnover in the kinds of persons who become zoo directors in this country within the past decade or so. Frequently, professionally trained people such as veterinarians or biologists, to mention two examples, have been recruited as directors. This is not always the case. I think it does not matter greatly which route one takes so long as the director is sympathetic to research and understands its place in the scheme of the modern zoo.

Communication and education, which go hand in hand, have been discussed a good deal. This is important from the standpoint of the research program's becoming established and accepted by a zoo, not only by the staff but by the board of trustees and/or the city council, depending upon how the zoo is financed. Equally important is the flow of information between institutions. There has been a tendency in the past to resolve a particular problem and guard it closely as the zoo's secret way to success. Dr. Van den bergh indicated that, when they published the diet developed for their hummingbirds at Antwerp some years ago, it was a new concept. Many wanted to keep the diet a closely guarded secret. The flow of information between institutions is much better than it used to be, but it can improve. There is so much left to be done in the time allotted to us. If we are to conserve this vast treasure of the animal kingdom now housed in the zoos throughout the world, communication with each other in solving common problems is urgently needed.

The researcher in a zoo must be tolerant. He must realize that he must fit into the scheme of things for, if he does not, the staff will soon be in disarray and people will be working at opposite ends. Or, perhaps even worse, "passive compliance" or no cooperation, will come to be. If the researcher's goals are known, almost without exception, attempts will be made by the technical staff to cooperate with him. He must understand, however, that he is not necessarily ushering the zoo into a bright new world that will solve all of its problems. Indeed, he is adding one additional problem to those that already exist, albeit a necessary one. The researcher cannot, as Peter Crowcroft stated at the beginning of our sessions, be someone who comes with an "empty shopping bag," wants to fill it, and then leave.

It is well to keep in mind that if research is undertaken in a zoo, serious consideration be given to the fact that research cannot interfere with the aesthetic value of the exhibits. If this is not done, the multiplicity of problems that can arise are sufficient to boggle one's mind and I think I need not dwell on them here. Acute or terminal experimentation, of course, could not be tolerated on animals used in exhibits. Recall that Dr. Dawes indicated that zoological parks cannot be expected to provide material for cancer research, although interesting pathologies may crop up in a collection of exotic creatures.

I have been asked to say something about the future of zoos. I do not know that I am the best qualified to speak to this particular point, but I have these thoughts that I would like to leave with you.

Yesterday we heard about a numbering system that would help in adding species of animals to a computerized census system. This was referred to as a taxonomic list. If my memory serves correctly, this was planned for 999 species. One thing that I hope zoos will do in the future, particularly in the United States, is to become more aware of invertebrate animals and exhibit them. As we probe the use of invertebrates in biomedical research and as knowledge expands concerning care and husbandry, these animals should be added to the list. The only other possible problem perhaps lies in the concept of subspecies. This can become confusing, for who can get systematists to agree as to what is a subspecies? If they cannot agree among themselves, how do you put it into a computer system? I would not worry about the subspecies unless there is a clear consensus among the systematists working with the group.

I look for closer communication between zoological parks and the biological and biomedical research communities. Surely, your own research programs will expand and, given the right opportunity, become more sophisticated.

One of the unique things that zoological parks have to offer the biomedical research community is the pool of expertise in handling exotic animals and an ability to provide new and exciting models for biomedical research. Were there sufficient time I would like to dwell on a few of these, but hopefully you will read about it another time.

It is my understanding that you will soon venture upon an accreditation system. I commend you heartily for having taken this step for I think it most essential. I heard someone say last night that they thought it was a mixed blessing, that guidelines are constrictive and would limit what could be undertaken. A peer evaluation is far better than one given on a routine basis by a bureaucratic official. Remember they are guidelines, they are not laws or rules cast in concrete. Such a system will help you provide far more than minimum care and therefore avoid a lot of the political and emotional pressures that can be brought against you. Developing the guidelines will be fraught with problems in the beginning and occasionally professional jealousy will rear its head, but in the long run I am sure you will reflect back on it as a highly important step and beneficial to your entire organization. Again, I commend you for it and I wish you the very best of success.

Research will gain broader acceptance; as this evolution takes place, it is my hope that a reproductive physiologist will be a standard appointment on a zoological park staff as pathologists and behaviorists are currently. In the vast array of the animals housed in zoos, we know almost nothing about the estrous cycles of these animals. Since it is going to be difficult to replace many of the stocks that we now have, time is of the essence. If we had studies concentrated upon the reproductive physiology of these animals, we then could perhaps bring about the induction of estrus in instances where animals have not been breeding. Techniques have been worked out for the analysis of progesterone. While it isn't possible for each zoo to buy an auto-analyzer, it can be done for you by hospital laboratories and commercial companies. Commercial laboratories have efficient technology and produce reliable results. Using them is far cheaper than investing in this kind of equipment, which also demands a skilled technician to tend it.

Lastly, I would like to leave you with some thoughts of what a zoo will be like 10, 15, or 20 years from now. You heard from the regulatory staff of the various bureaus Monday morning that, within 5 years, no animal would be coming into the United States without a permit. As the ever-expanding human populations of the world compete for protein sources, staple items of food that have been routine in zoological parks will undoubtedly be sought as human fare. Likely, the broad coverage of mammalians and bird populations within a particular zoo won't be as great 20 years from now unless we solve some of the problems that we have been unable to resolve in the past. It is entirely possible that a particular zoological park will specialize on one area of the world or a select group of animals.

Likely, zoos will be concerned more with animals closer to them. This is a point that should not be overlooked. Many zoological parks are located in areas where there are endemic, but rare, species. Were they exotic species they would be sought with avarice by zoos throughout the country. Some of them will be challenging models for research. As an example, the mountain beaver of the genus *Aplodontia* is a relict species of animal that occurs in the Pacific Northwest. It has four species of Siponaptera, three of which are so unique that subfamilies have had to be erected for them. The fourth, which belongs to a rather common genus, is the largest flea known in the world. If the ectoparasites can be this different, the mountain beaver ought to have something apart from modern mammals and could provide us with some valuable insights that we lack. It is well to remember that our native fauna is just as interesting and as worthy of saving for posterity as the exotic animals.

It has been a privilege to be with you and it is hoped that sometime in the not too distant future we can undertake a joint program of this nature again. I thank you.